Clusters in a Cold Climate

The Innovation Systems Research Series

Innovation, Institutions and Territory: Regional Innovation Systems in Canada, J. Adam Holbrook and David A. Wolfe, editors

Knowledge, Clusters and Regional Innovation: Economic Development in Canada, J. Adam Holbrook and David A. Wolfe, editors

Clusters Old and New: The Transition to a Knowledge Economy in Canada's Regions, David A. Wolfe, editor

Clusters in a Cold Climate: Innovation Dynamics in a Diverse Economy, David A. Wolfe and Matthew Lucas, editors

Related Publication

Understanding Innovation in Canadian Industry, Fred Gault, editor

Clusters in a Cold Climate

Innovation Dynamics in a Diverse Economy

Edited by David A. Wolfe and Matthew Lucas

Published for the School of Policy Studies, Queen's University
by McGill-Queen's University Press
Montreal & Kingston • London • Ithaca

National Library of Canada Cataloguing in Publication

Innovation Systems Research Network. Conference (5th : 2003 : Ottawa, Ont.)
Clusters in a cold climate : innovation dynamics in a diverse economy / edited by David A. Wolfe and Matthew Lucas.

(The innovation systems research series ; 4)
Papers presented at the 5th Annual Conference of the Innovation Systems Research Network, held in Ottawa, Ont., May 1-2, 2003.
Includes bibliographical references.
ISBN 1-55339-039-3 (bound).—ISBN 1-55339-038-5 (pbk.)

1. Industrial location—Canada—Congresses. 2. Technological innovations—Economic aspects—Canada—Congresses. I. Wolfe, David A. II. Lucas, Matthew III. Queen's University (Kingston, Ont.). School of Policy Studies IV. Title. V. Series: Innovation systems research series ; 4.

HC120.I55I55 2003 338.6'042'0971 C2004-902039-0

This book is dedicated to Lisa and Michaella for their constant support and encouragement.

Table of Contents

Tables and Figures

Tables

FIGURES

1

Introduction: Clusters in a Cold Climate

David A. Wolfe and Matthew Lucas

The title of the present volume from members of the Innovation Systems Research Network seems singularly appropriate as Canada emerges from one of the coldest winters in the past decade and a half. The reflexive reference is not to our recent weather, but rather to an early premonitory report from the Science Council of Canada. In its report of 1971, *Innovation in a Cold Climate*, the Science Council warned that science-based industries in Canada were deteriorating and that this deterioration jeopardized the country's ability to sustain its recent record of substantial investments in education and social welfare. It concluded that an improved capacity for technological innovation with an increased use of skilled labour was crucial to the ability to meet rising social and economic expectations (Science Council of Canada 1971).

While the position of Canadian manufacturing, and particularly its high technology sectors, has improved dramatically in the intervening three decades, the themes articulated in this report continue to populate recent policy documents and statements from the senior levels of government. Both our investments in scientific and technological infrastructure and our overall level of performance with respect to research, development, and innovation have increased significantly since the early 1970s. Yet, our performance relative to the other leading industrial countries, especially the United States, has failed to improve to a commensurable degree.[1] The long-term implications of this lack of improvement for our standard of living and our ability to fund education and social welfare programs continue to concern policymakers. However, the ability to translate this concern into specific policy actions has been impeded by disagreement about the main reason for our lagging innovative performance and the most appropriate policy tools to remedy the problem.

On various occasions in the past several years, a more innovative economy, an increase in the role of dynamic industrial clusters, and a greater ability to commercialize our investments in scientific research have all been identified as important to overcoming this problem. However, the precise relationship between these factors has not always been as clear. In a widely quoted speech to the Montreal Board of Trade in September 2003, the Liberal leadership candidate, now prime minister, Paul Martin, suggested that three essential elements were necessary to create the conditions for sustainable long-term growth in the twenty-first century: continuing investment in university-based research by both the public and the private sectors; better support for small and medium-sized enterprises to access the resources needed to innovate and develop new products and to implement best practice technology to produce those products; and finally, better access to capital to enable both large and small firms to innovate and develop these products. These themes were broadly echoed in the recent Speech from the Throne and the following Address by the Prime Minister. In the Throne Speech, the government declared that one of its key goals is building a strong, innovative economy for the twenty-first century as a prerequisite for also building a fairer and more equitable society. The government's goal is to make Canada a world leader in developing and applying path-breaking technologies. However, innovative firms in Canada face two key obstacles that impede their ability to deliver on this potential — access to early-stage financing and the capacity to commercialize new ideas and grow their business — both of which the government committed itself to remedy (Canada. Governor General 2004).

These new priorities outlined in both the Speech from the Throne and the Prime Minister's reply indicate a subtle but important shift away from some of the language and concepts employed in recent years, especially as reflected in the Innovation Agenda of 2003 and the extensive consultations that it entailed.[2] This shift may reflect, in part, the changing priorities of a new government, but it may also reveal dissatisfaction with the outcome of the previous consultation exercise or a feeling that the agenda was too broad to be readily translated into concrete policy initiatives. The danger is that the new approach risks discarding some of the insights into how and why firms innovate that had been acquired through the previous policy consultations and through the broad body of policy-relevant research that has been accumulating in this country over the past decade. Of equal concern is the possibility that this approach signals a return to the single instrument, top-down system of delivering policy solutions that has traditionally characterized much of the federal government's (and the provinces' for that matter) approach to this field.

If a top-down approach recurs, what will be lost is the fundamental insight acquired through the innovation systems approach, developed by researchers in Canada and abroad, that a firm innovates in a social context as part of a broad network of related firms and institutions that all contribute to its innovative capability. The extensive body of research undertaken since 1997 by members of the Innovation Systems Research Network (ISRN) in Canada on the role of innovation systems, and the related concept of industrial clusters, offers a strong degree of empirical evidence to explain the nature of these complex and multi-faceted relationships across a broad cross-section of regions and industries in this country. This research helps us understand the way in which the innovative capabilities of firms are fundamentally grounded in the networks and institutions that compose their regional innovation systems or industrial clusters (de la Mothe and Paquet 1998; de la Mothe and Niosi 2000; Holbrook and Wolfe 2000, 2002; Wolfe 2003).

The chapters contained in this volume present the latest research findings from the ISRN's multi-year research initiative into the dynamics of industrial clusters and regional innovation systems. The findings expand our understanding of industry clustering in Canada with the latest results of eight new case studies across a sample of both established and emerging industries: auto parts in southern Ontario, wine in the Niagara region, biotechnology in Vancouver and Halifax, optics/photonics in Quebec City, and multimedia in Vancouver, Montreal, and Toronto. These industries vary considerably in terms of the average size and age of firms, the type and frequency of inter-firm linkages, the patterns of labour organization, the intensity of research and development, and the economic impact on the surrounding region. The case studies are therefore valuable both for the insights they offer into particular industries and regions and for the opportunities they afford for comparing cluster dynamics across a wide range of industry characteristics and regulatory environments. They should also help us identify dynamics that may be unique to the Canadian context.

For each concentration of firms and supporting organizations the authors attempt to answer the question: Is it a viable cluster? That is, does a particular collocation of firms, financial and legal services, research and training institutes, business and civic associations, and public supports promote sustainable endogenous innovation and industry growth, and if so, how? The question has a counterpart at the level of individual firms: Does locating within a cluster help a firm access resources and markets, deal with economic uncertainty, and adapt to market changes? There are three steps to answering these questions: we must first identify actors that provide key inputs to the industry (such as knowledge, technology, finance, and labour), actors that serve as markets for key outputs, and actors that facilitate interaction within the region. We must then identify linkages between these actors including both formal participation in supply chains, research consortia, and civic associations, and informal relationships established through shared employment, educational histories, and personal exchanges between key individuals. Finally, we must measure how these interactions affect the capacity of individual firms and the industry as a whole to remain locally vibrant and globally competitive. This final step is perhaps the most difficult as it requires us to identify intangible factors such as social capital and culture that sustain a creative and entrepreneurial business environment.

In carrying out this process the various members of the research team faced three challenges unique to the Canadian context and to the particular industries under investigation: the overwhelming influence of the nearby US market, the youth and instability of many of the clusters, and the lasting effects of the recent economic downturn in the telecommunications industry. All of the clusters discussed here, and indeed most Canadian firms, are located within a short drive of the large and highly competitive US market, a market that can distort our perception of local cluster dynamics. Most of the case studies included in this volume generate a significant portion of their revenues from exports, primarily to the US, which means that the most important supply-chain relationships often extend beyond the region. These external relationships can easily overshadow linkages with local organizations that sustain information and labour flows within an industry, and lead to the simplistic conclusion that global ties are more important than local ties. Moreover, three of the five industries examined here, biotechnology, multimedia, and optics/photonics, are young and rapidly evolving and have yet to stabilize cluster relationships. It is difficult to determine if the absence of strong inter-firm linkages indicates that a cluster is weak or that it is immature. This has direct policy implications, for if the cluster is not viable then efforts to support it should be redirected to more successful industries; but if the cluster is merely at an early stage of development, one could argue that supports should be intensified. The recent dramatic downturn in the telecommunications industry also makes it difficult to judge a cluster's viability. The multimedia industry in particular experienced radical restructuring in recent years and is still adjusting to the new economic environment. Despite these challenges our authors highlight crucial strengths and weaknesses in each of the clusters and make well-reasoned estimates of each cluster's viability.

Overall the studies accentuate some of the key themes identified in the previous research output from members of the ISRN. The picture emerging from our study departs substantially from much of the received wisdom on the role and nature of industrial clusters — most notably concerning the alleged importance of a strong local customer base and strong local competition in spurring the emergence and evolution of dynamic, knowledge-based clusters. Nor is it evident from our findings that direct, non-market interaction and knowledge-sharing between local firms in the same industry is rampant. Our evidence suggests that where such interaction occurs, it is indirect and mediated through civic associations and other local organizations. While this form of local learning is considerably more prevalent between firms and their local suppliers, not all inputs are locally sourced. In particular, it appears that a large component of the knowledge inputs to local production, at least in certain sectors, is drawn from well outside the region. Furthermore, in posing the question of what anchors these firms in their present location, the answer that repeatedly emerges from many of the studies is the depth and quality of the local labour market and the role of institutions of higher education in identifying and responding to the needs of clusters (Wolfe and Gertler 2004). These findings bear important policy implications, suggesting that government efforts to support innovation and commercialization may be targeted more effectively at groups of firms through networks and industrial clusters rather than on an individual basis.

The volume begins with an examination by Fitzgibbon, Holmes, Rutherford and Kumar of the auto parts industry in southern Ontario, specifically the part of the industry located in the two regions of Windsor-Essex County and Kitchener-Waterloo-Cambridge-Guelph. A "traditional" industry of vital economic importance to the region and province, auto parts manufacturing is inherently linked to the auto-assembly industry. The authors note the existence of strong vertical interdependencies between local firms that are part of the long-standing multi-tiered supply chains feeding the auto-assemblers. Among other insights, this study reveals the vulnerability of a well-established cluster to economic and regulatory changes outside the industry's and the region's control. The interdependencies between assemblers and parts manufacturers have been threatened by dramatic shifts within the auto-assembly industry driven by the World Trade Organization rulings against the Canada-US Auto Pact, the opening up of the North American markets to European and Japanese assemblers, the proliferation of financial incentives to relocate plants to the southern US, and the recent rise in the Canadian dollar. Facing an increasingly competitive marketplace, assemblers have demanded that their parts suppliers cut costs and increase productivity. In response, suppliers have had to increase their innovative capacity, which they have done in part by increasing investments in research and development. Small firms unable to support these additional expenses may disappear. This evidence of creative destruction at work clearly demonstrates that a long history of inter-firm cooperation is not enough to preserve a cluster when the competitive environment shifts, although it may help the cluster adapt more quickly. This case also alerts us to the danger of relying on exogenous innovation to feed product development. The success of recent efforts to develop local research and development in auto parts may determine the future of the auto parts industry in Canada.

Industry evolution, or the failure to evolve, is also an important issue in the Niagara wine cluster. Mytelka and Goertzen focus their analysis on two prominent institutions, the Ontario Grape Growers Marketing Board (OGGMB), now the Grape Growers of Ontario, and the Liquor Control Board of Ontario (LCBO). Both played an important role in the dramatic transformation of Niagara's wine industry in the late 1980s and early 1990s by encouraging and facilitating the introduction of new vines and vintner techniques that significantly improved wine quality. The two also oversaw an initial consolidation of the wine industry into a small number of large wineries, a subsequent growth in the number of small growers and vintners, and a trend toward grower-vintner establishments that combined both operations. Although this last shift changed the dynamic between growers and vintners, the OGGMB continued to treat the two as separate entities with contrary interests. Mytelka and Goertzen argue that the OGGMB's inability to adapt to the new environment may undermine its potential contribution to the development of the cluster. In contrast, the LCBO has evolved considerably over the past two decades, shifting its regulatory and sales responsibilities from a focus on nurturing a struggling industry to a focus on maximizing profits. As the world's largest purchaser of wine, the LCBO controls the domestic market for Ontario vintners. A decision to focus sales on high turnover stock has reduced the shelf space available to the many small wineries in the province. While noting the steps the LCBO has made to correct this problem, Mytelka and Goertzen argue that the organization should revive its early model of supporting the cluster as a whole.

We now turn from mature industries to relatively immature ones. Rosson and McLarney's analysis of the biotechnology cluster in and around Halifax, and Holbrook, Salazar, Crowden, Reibling, Warfield and Weiner's analysis of the biotechnology cluster in the Vancouver area, both point to the challenges that specialized industries face in small domestic markets. Firms in the Halifax region in particular are more closely connected to customers, suppliers, and competitors in the US than within the region. Both studies highlight the important role that large public research institutes, particularly universities, played in creating and incubating the clusters. An impediment to cluster growth in both regions is a lack of specialized legal and financial services. The clusters are presently too small to sustain these supports. Rosson and McLarney conclude that while the collection of firms does not yet constitute a viable cluster, it is too early to rule out future success. Holbrook *et al.* note that the Vancouver industry is based on the production and sale of intellectual property, rather than physical products, to firms outside the region. As a result there are few supply chain relationships within the cluster. This raises the question of whether this is a viable approach to industry growth or whether it indicates systemic weaknesses in the region. It also suggests the difficulty of growing a young cluster close to a large competitive market that is able to acquire and relocate small potentially valuable firms. Whether the Vancouver and Halifax biotechnology industries can make the transition to sustained growth remains to be seen.

Another research-intensive cluster is the optics/photonics industry in and around Quebec City, which creates highly specialized products for niche markets, mostly in the US. Measuring the frequency and strength of inter-firm linkages within the region, Keroack, Ouimet and Landry conclude that weak links between firms are more prominent than strong links. They also note the central role of large public sector research institutes both in giving birth to the cluster and sustaining knowledge flows through it. They suggest that such institutes may play a particularly important role in Canada because of the low private investment in research and development. The optics/photonics industry, much like the biotechnology industry, relies on global linkages for key knowledge inputs as much as it relies on local linkages to absorb and apply that knowledge. Keroack *et al.* also suggest that servicing niche global markets may be a fruitful strategy for industries with limited local demand, although this does increase the danger that disruptions in those niche markets or the entry of new competitors could devastate the local industry.

From an industry producing niche products we turn to an industry producing such a wide range of products and services that the industry itself is difficult to classify. All of the authors investigating the multimedia industry—Smith, McCarthy and Petrusevich on the Vancouver cluster; Tremblay, Chevrier and Rousseau on the Montreal cluster; and Britton and Legare on the Toronto cluster — address the challenge of defining an industry that cuts across a number of traditional sectors. While the industry differs somewhat among the three regions, there are a number of important similarities. All note that multimedia emerged in the mid-1990s when the confluence of content and computing resulted in the digitization of images, sounds, text, and video and their creative presentation in a wide range of increasingly interactive products. The industry's enduring value lies in the skills and creativity of its labour force. In all three clusters the industry is young, evolving rapidly, and populated by a large number of small firms. Firms contract much of their work on a project basis and therefore have irregular revenues and labour requirements. While some firms collaborate on proposals, particularly those serving the business market in Toronto, most subcontract work to freelance workers when they are unable to perform it in-house. The authors note that the small size and limited resources of individual firms impede their ability to interact with other firms, support organizations, or government programs. The flow of information and skills through the industry depends on its highly mobile labour force and the practice of subcontracting work to freelancers, which moves workers through the industry. In the wake of the dot-com crash in the late 1990s and subsequent downturn in the telecommunications industry, there has been a high turnover of employees and firms within the industry. This rapidly changing environment creates a number of challenges for cluster growth as inter-firm linkages are difficult to maintain. Similar to the biotechnology industry, the multimedia industry's largest and most demanding customers are outside the region and international sales are seen as a way of building one's local reputation.

The detailed analysis of the diverse array of Canada's industrial clusters presented in this volume provides some important policy lessons for those responsible for crafting the country's next generation of innovation and commercialization policies. In the first instance, the industrial clusters across the country are characterized by a high degree of diversity both across industrial sectors and across regions within the same sector. In each instance the clusters under examination are confronted by a comparable set of problems. From established clusters in auto parts and wine to emerging clusters in biotechnology, photonics and multimedia, firms face the challenges of learning how to adapt to rapidly changing demand and cost conditions, increased competition both at home and abroad, and new disruptive technologies. However, the ways in which these factors impact individual clusters vary considerably and appropriate policy responses must be tailored to reflect this variation. At the same time, the studies underline the salience of a number of key themes emerging from the research: the contribution of both local and global sources of knowledge to innovation, the powerful stimulus of external markets on innovative behaviour, the interaction between key elements of the research infrastructure and innovative firms, and the complex but essential role that various aspects of public policy, especially that concerning research and teaching in the higher education sector, play in influencing cluster dynamics.

Appropriate public policy must take these dynamics into account if it is to achieve its maximum positive impact. The wave of innovation policies implemented in the past decade has created a dense network of research institutions and technological infrastructure that has greatly strengthened Canada's research capabilities. An increased emphasis on research-industry linkages has also improved knowledge flows within the regional innovation system. On the downside, it is now virtually impossible for bureaucrats, let alone private firms, to track the plethora of policies and programs. A key challenge for innovation policy is to ensure a better integration and coordination of programs and policy instruments. As the ISRN's continuing research program suggests, this can most often be accomplished at the local and regional level, from the perspective of strategic clusters or local and regional innovation systems.

Many existing policies and programs have been implemented in a traditional fashion, administered by individual departments or agencies with little cross-jurisdictional coordination and often little attention paid to the broader implications of the program for cluster development in the local or regional context. We need a more integrated and joined-up approach to policy planning at the "governance" level that builds on existing program capabilities. There is a growing recognition that economic development policies work most effectively when the direct beneficiaries of those policies and programs play a key role in both their design and implementation. This involves developing a rolling set of innovation strategies at the cluster, local and regional levels to ensure that the existing R&D infrastructure and economic development programs are used to maximum advantage — to assess existing needs and identify gaps in the program array (Gertler and Wolfe 2004). The time has come to build on insights derived from the recent empirical research and policy analysis to advance the pace of social learning in this crucial area.

NOTES

[1]For a more detailed review of recent policy statements and the concerns expressed over Canada's R&D performance, see Wolfe (2002).

[2]For a more extensive review of the Innovation Agenda and the associated consultations, see de la Mothe (2003).

REFERENCES

Canada. Governor General. 2004. "Speech from the Throne. To Open the Third Session of the Thirty-seventh Parliament of Canada." Ottawa, 2 February.

de la Mothe, J. 2003. "Ottawa's Imaginary Innovation Strategy: Progress or Drift?" in *How Ottawa Spends 2003–2004: Regime Change and Policy Shift*, ed. G.B. Doern. Toronto: Oxford University Press.

de la Mothe, J. and G. Paquet, eds. 1998. *Local and Regional Systems of Innovation.* Amsterdam: Kluwer Academic Publishers.

de la Mothe, J. and J. Niosi, eds. 2000. *The Economic and Social Dynamics of Biotechnology.* Dordrecht, Netherlands: Kluwer Academic Publishers.

Gertler, M.S. and D.A. Wolfe. 2004. "Local Social Knowledge Management: Community Actors, Institutions and Multilevel Governance in Regional Foresight Exercises," *Futures* 36 (February): 45-65.

Holbrook, A. and D.A. Wolfe, eds. 2000. *Innovation, Institutions and Territory: Regional Innovation Systems in Canada.* Kingston and Montreal: School of Policy Studies, Queen's University and McGill-Queen's University Press.

—— 2002. *Knowledge, Clusters and Regional Innovation: Economic Development in Canada.* Montreal and Kingston: School of Policy Studies, Queen's University and McGill-Queen's University Press.

Science Council of Canada. 1971. *Innovation in a Cold Climate: The Dilemma of Canadian Manufacturing.* Report No. 15. Ottawa: Information Canada.

Wolfe, D.A. 2002. "Innovation Policy for the Knowledge-Based Economy," in *How Ottawa Spends 2002–2003: The Security Aftermath and National Priorities,* ed. G.B. Doern. Toronto: Oxford University Press.

—— , ed. 2003. *Clusters Old and New: The Transition to a Knowledge Economy in Canada's Regions.* Kingston and Montreal: School of Policy Studies, Queen's University and McGill-Queen's University Press.

Wolfe, D.A. and M.S. Gertler. 2004. "Clusters from the Inside and Out: Local Dynamics and Global Linkages," *Urban Studies* 41 (May).

2

Shifting Gears: Restructuring and Innovation in the Ontario Automotive Parts Industry

Susan Fitzgibbon, John Holmes, Tod Rutherford and Pradeep Kumar

Introduction

> For two decades, the bland landscape between Toronto and Windsor has been home to a great wealth-creation machine: Canada's auto players ... An industry [the auto parts industry] that looks grubby and quaintly old-fashioned from the outside has in reality been a hotbed of job growth, innovation, and capital spending. Its success has allowed Canada to skirt recession as other industries implode (Reguly 2003).

Together with the motor vehicle-assembly industry, the automotive parts industry in Canada has expanded significantly over the last two decades, especially during the 1990s. The total value of manufacturing shipments in the industry rose over seven-fold between 1980 and 1999; labour productivity, measured by value-added in manufacturing per hour paid, more than tripled and employment doubled.[1] Over 90 percent of employment in Canada's automotive industry is concentrated in the broad belt between Windsor and Oshawa, Ontario. There is little exaggeration, therefore, in the claim that the economic prosperity enjoyed by southern Ontario during the 1990s owed at least as much to the booming automotive industry as it did to the rise of "new" industries such as biotechnology, photonics, and new media. The automotive industry is often caricatured as an "old-fashioned metal bashing industry." Nothing could be further from the truth. Substantial new investment, the development and uptake of new technologies such as new materials, electronics, and informatics, and the continued pioneering of new advanced manufacturing methods all suggest that the automotive industry is just as important in the so-called New Economy as it was in the old.

The automotive industry plays a major role in the Ontario economy. In 2002, the $95 billion in value of shipments produced by the industry accounted for nearly 20 percent of the province's manufacturing gross domestic product (GDP). With about 50,000 workers in the vehicle-assembly industry and a further 100,000 in automotive parts manufacturing, the province was the eighth largest producer of motor vehicles in the world (Ontario. MEOI 2003). Over the last decade the industry attracted almost $20 billion in investment, and in 2002 investment commitments in the industry exceeded $2 billion, making Ontario the third highest ranked jurisdiction with regard to auto investment (APMA 2002). Industry experts, however, caution that despite its recent success the industry currently faces a number of significant challenges.

The "automotive cluster" in Ontario is highly significant within the North American automotive industry. Researchers at the Institute for Competitiveness and Prosperity recently took 41 clusters that Michael Porter had identified in the United States and mapped them onto Ontario. The automotive cluster in Ontario ranked considerably higher than it did for the US economy as a whole.[2] At the provincial and state level, Ontario ranked third, behind only Michigan and Ohio, and at the Census Metropolitan Area (CMA) level, Toronto, Oshawa and Windsor ranked second, eleventh, and fourteenth respectively (Institute for Competitiveness and Prosperity 2002, pp. 35-38).

Our study focuses on the automotive parts industry in southern Ontario. Many of the other studies being undertaken under the ISRN project focus on industries geographically concentrated in tightly circumscribed regions or particular communities. While the Canadian automotive parts industry is heavily concentrated in southern Ontario, it stretches over several hundred kilometres and is a significant presence in a number of widely separated communities including the Greater Toronto Area (GTA), Windsor, Oshawa, Kitchener-Waterloo, London, Guelph, Stratford, Brantford, and St. Catharines (see Figure 1). Furthermore, while the automotive industry in North America (Canada, the United States, and Mexico) is highly integrated across the

FIGURE 1
Automotive Plants, Southern Ontario, 2002

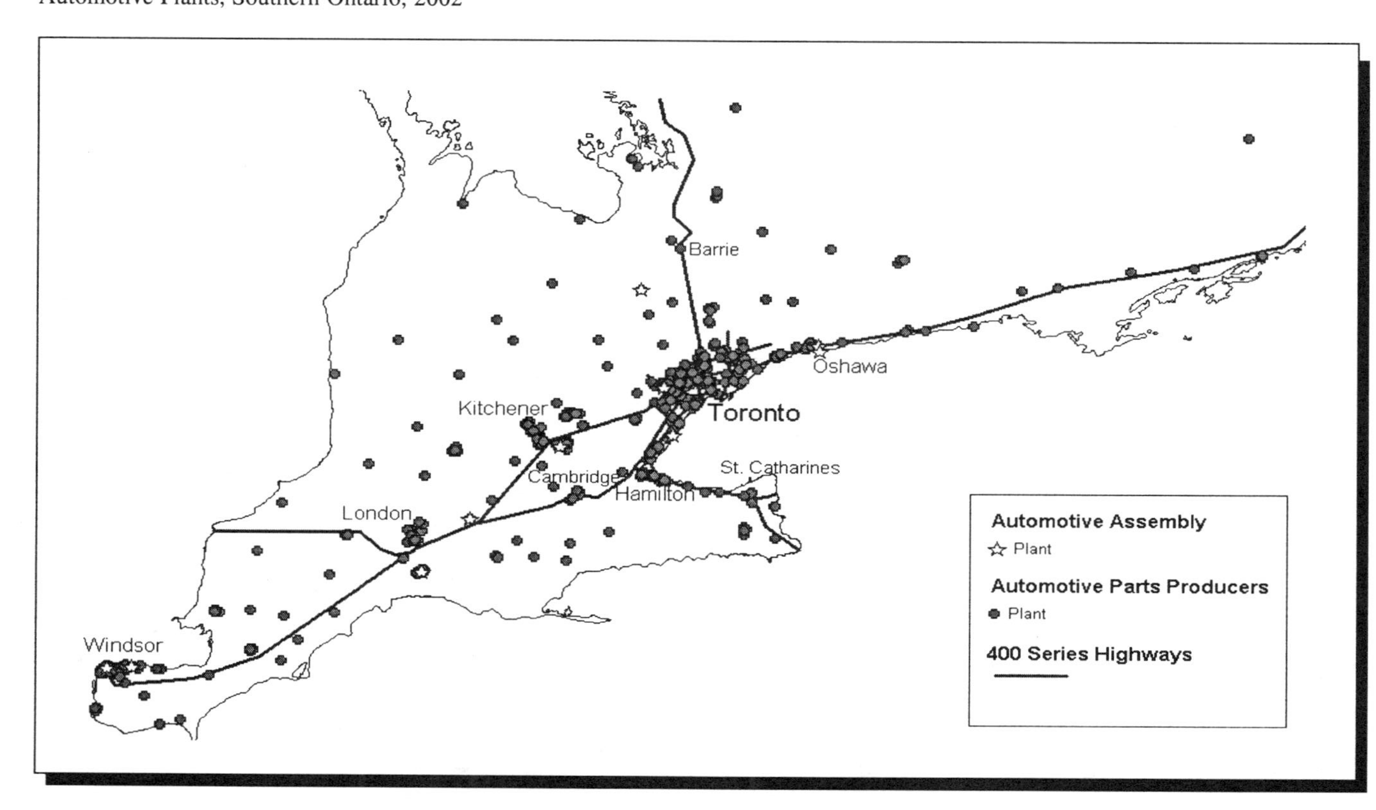

continent, there are several distinct regional production concentrations. Ontario is part of one such concentration centred on Detroit and spanning the Canada-US border. The concentration includes southern Ontario and the states of Michigan, Ohio, Illinois, Indiana, Kentucky, and Tennessee. Thus, it is possible to identify geographical "clusters" in the automotive parts industry at a number of different scales ranging from clusters in particular Ontario communities, through the broader concentration of auto parts plants in the Windsor to Oshawa section of the province, to the much larger transborder geographical concentration centred around the lower Great Lakes of which southern Ontario is a part.

The size and geographical extent of the industry posed particular challenges for the research design of our project. Our study is being conducted on two different levels. At one level, we examine the structure, performance, and institutional setting of the industry within southern Ontario as a whole. This analysis relies heavily on secondary data from Statistics Canada and interviews with key actors in the industry.[3] At a second level, the intensive interviewing phase of our study, we focus on two of the sub-regional automotive parts clusters in Ontario; namely, the Windsor-Essex County area and the Kitchener-Waterloo-Cambridge-Guelph area. Windsor was chosen because of its long-standing involvement in the auto industry dating back to the first decade of the twentieth century. As Desrosiers observes, "Windsor-Essex County is uniquely positioned within the automotive manufacturing belt of southern Ontario. Its location directly adjacent to Detroit gives local automotive suppliers unique access to original equipment manufacturers (OEM) decisionmakers. Many industry observers and executives believe that Windsor-Essex County is the only viable Canadian location for automotive research and development" (n.d., p. ix). The Kitchener-Waterloo-Cambridge-Guelph sub-regional concentration was chosen because, in addition to being home to a significant number of firms involved in the automotive parts industry, it is also the subject of another ISRN study focused on the wireless and photonics sector. This will enable comparisons to be made between the innovative capacities of two very different industries located in the same region.

In this, the first chapter from the ISRN automotive parts industry study, we focus on the structure and recent performance of the industry and discuss the major challenges it now faces.

SIZE AND STRUCTURE OF THE AUTOMOTIVE PARTS INDUSTRY IN CANADA: AN OVERVIEW

Defining the Industry

An all inclusive definition of the automotive parts industry is difficult due to the number and diversity of components that go into the assembly of motor vehicles. The North American Industrial Classification System (NAICS) used for the *Annual Survey of Manufactures* (ASM) conducted by Statistics Canada divides automobile manufacturing into two major industries at the four-digit classification level: motor vehicle manufacturing (NAICS 3361) and motor vehicle parts manufacturing (NAICS 3363).[4] The latter is further divided at the five-digit level into eight sub-industries (Table 1).[5] Several observations can be made regarding this official classificatory definition. First, rather than being one coherent and homogeneous industry, the automotive parts industry consists of a number of distinct sub-industries which simply share a common set of customers. The various sub-industries exhibit significantly different industrial organizational characteristics such as capital intensity, degree of technological complexity of products, and skill requirements. As a result, an analysis that relies only on data at the NAICS four-digit level can mask quite different circumstances and competitive dynamics at the sub-industry level. Second, establishments manufacturing products that one normally thinks of as being auto parts — tires, batteries, hoses and belts, for example — are classified in industries that lie outside NAICS 3363. Similarly, we have come to appreciate that a very significant proportion of the machine, tool, die and industrial mould (MTDM) industry in the Windsor-Essex County region is dedicated to the automotive industry and should rightly be considered to be part of the automotive cluster in Windsor.[6]

Third, in sub-industries such as metal stamping and plastic parts, some establishments simultaneously may be supplying customers in other industries as well as the auto industry and, over time, may move in and out of supplying the auto industry. In other words, the "boundary" between the automotive parts industry and other industries is constantly in flux.

As noted earlier, our study focuses on the automotive parts industry in southern Ontario and in particular on the two sub-regions of Windsor-Essex County and Kitchener-Waterloo-Cambridge-Guelph. Given the industry's

TABLE 1
Automotive Parts Sub-Industries, Canada and Ontario, 2001

Industry/ Sub-industry	*NAICS Code*	*Products*	*Establishment Count (% of total)*		*Total Number of Employees (% of total)*		*Value of Shipments ($ 000,000) (% of total)*	
			Can	*Ont*	*Can*	*Ont*	*Can*	*Ont*
Engine and Engine Parts	33631	Carburetors/ fuel injectors, cylinder heads, manifolds, camshafts	162 (15.5)	81 (12.0)	11,110 (10.1)	9,811 (9.9)	5,410 (16.8)	5,249 (17.1)
Electric and Electronic	33632	Alternators, instrument boards, fuel pumps, lighting fixtures, wiring assemblies	120 (11.5)	74 (10.9)	6,886 (6.2)	5,965 (6.0)	1,554 (4.8)	1,449 (4.7)
Steering and Suspension	33633	Struts, shock absorbers, power steering pumps, tie rod assemblies	43 (4.1)	30 (4.4)	4,912 (4.4)	4,811 (4.9)	1,377 (4.3)	1,364 (4.4)
Brake Systems	33634	Master/wheel cylinders, drums and rotors, calipers, vacuum boosters	64 (6.1)	41 (6.1)	6,840 (6.2)	5,551 (5.6)	1,523 (4.7)	1,336 (4.4)
Transmission and Power Train	33635	Rear axle assemblies, transmissions and parts, clutch assemblies	86 (8.2)	51 (7.5)	10,930 (9.9)	10,012 (10.1)	3,631 (11.3)	3,504 (11.4)
Seating and Interior	33636	Seat frames/seats, interior panels, trim, seat belts, visors	81 (7.7)	53 (7.8)	10,589 (9.6)	10,117 (10.2)	3,681 (11.4)	3,602 (11.7)
Metal Stamping	33637	Body panels, firewalls, floor pans, hubcaps	103 (9.8)	88 (13.0)	15,525 (14.1)	14,481 (14.6)	4,235 (13.1)	4,053 (13.2)

TABLE 1
(Continued)

Industry/ Sub-industry	*NAICS Code*	*Products*	*Establishment Count (% of total)*		*Total Number of Employees (% of total)*		*Value of Shipments ($ 000,000) (% of total)*	
			Can	*Ont*	*Can*	*Ont*	*Can*	*Ont*
Other Parts	33639	Air bag assemblies, exhaust systems, air-conditioning systems, wheels	238 (22.7)	153 (22.6)	25,249 (22.9)	22,412 (22.6)	7,176 (22.2)	6,745 (22.0)
Motor Vehicle Parts	3363	See above	897	571	92,041	83,160	28,588	27,302
Automotive Plastic Parts	326193	Fiberglass bodies, mouldings and extrusions, lenses, trim pieces	150 (14.3)	105 (15.5)	18,374 (16.6)	16,033 (16.2)	3,672 (11.4)	3,384 (11.0)
Motor Vehicle Parts and Automotive Plastic Parts	3363 + 326193	See above	1,047	676	110,415	99,193	32,259	30,685

Source: Statistics Canada (2004*b*).

exceedingly high degree of geographical concentration in southern Ontario, ASM data for the Canadian industry as a whole serves as a reasonable surrogate measure for the industry in Ontario (Table 1).[7] Leaving aside the miscellaneous category of "other parts," metal stamping, engines and engine parts, and plastic parts stand out as the three most important automotive parts sub-industries in both Canada and Ontario with regard to employment and value of output. These three sub-industries are well represented in both of the sub-regional concentrations on which our study will focus. Windsor has a broad range of automotive parts plants tied to assembly plants in Windsor and nearby Detroit. Ford has four large parts plants in the city — two engine plants and two aluminum casting plants — and General Motors (GM) operates a transmission plant. Metal stamping has always been important, and, as noted earlier, Windsor has a world-class industrial mould-making industry tied almost exclusively to the automotive industry.[8] In the Kitchener-Waterloo-Cambridge-Guelph region metal stamping and machined engine and transmission parts predominate.

The Structure of the Industry: Assembler-Supplier Relationships

Although highly interdependent, the motor vehicle-assembly and automotive parts industries have markedly different industrial structures. The assembly industry in Canada, which comprises 12 high-volume light vehicle-assembly plants all located between Windsor and Oshawa in southern Ontario, is dominated by a handful of very large foreign-owned global firms.[9] GM, Ford, and DaimlerChrysler (known collectively as the "Big Three") account for 75 percent of light-vehicle production and employment in vehicle assembly in Canada (CAW 2003). The remaining 25 percent is shared between CAMI (a joint venture between GM and Suzuki), Honda, and Toyota.

In sharp contrast, the automotive parts industry is characterized by a large and diverse collection of firms. These range from companies operating a single plant and employing fewer than 30 workers to large multinational corporations employing several thousands of workers spread across a number of plants. However, as one might expect, employment and production are highly concentrated in the larger establishments with over 200 employees. The vast majority of these larger plants are operated by one of a score of large manufacturers, a significant number of which are Canadian owned. In a recent report, Pilorusso (2002) divided the automotive parts industry in Canada into three principal segments:

- A group of globally competitive Canadian companies which account for about one-third of employment and output in the industry and represent the segment of the industry that grew most rapidly during the 1990s;
- A substantial number of plants owned by foreign globally competitive component manufacturers that account for approximately 50 percent of employment and output; and
- A large number of small Canadian-owned plants and a small number of foreign-owned plants in transition that account for the remaining 20 percent of employment.

Since 1990, a combination of changes in relationships between suppliers and assemblers and the increasing global reach of assemblers have had a direct impact on the global automotive parts industry in general and specific ramifications for the structure of the automotive parts industry in Canada. In the late 1980s and early 1990s, lean production in the automobile industry focused on reducing costs within assembly plants through the reduction of waste, the reorganization of work, and the increased outsourcing of parts production. In the late 1990s, the focus shifted from the assembly plants to achieving cost reductions across the entire supply chain. This has led to recent and ongoing changes in assembler-supplier relationships that, in turn, have modified the structure of the automotive parts industry (Flynn and Belzowski 1996; Holmes 2004*a*).

Original equipment manufacturers have forced the development of distinct tiers among their suppliers and drastically reduced the number of suppliers delivering directly to OEM assembly plants (so-called first-tier or Tier 1 suppliers) to cut the time and cost involved in managing the supply chain. At the same time, OEMs have increasingly shifted the responsibility for pre-production design and engineering activities to their suppliers. First-tier suppliers are moving from being systems suppliers to system integrators supplying modules such as complete instrument panels, seating/interior trim systems, climate control systems, and door systems. Increasingly, the first-tier supplier is responsible for not only the assembly of parts into modules but also for the management of lower-tier suppliers providing the parts. The OEMs are becoming more and more involved in the specification and tight monitoring of the production and quality systems employed by their first-tier suppliers and are pressing their first-tier suppliers to institute similar practices with regard to the latter's own suppliers. The overall effect is a progressive standardization of processes from the OEM on down the supply chain.[10] The assemblers are standardizing platforms and models across their global divisions to reduce development costs and obtain economies of scale. One result is a strong desire on the part of the OEMs to use the same first-tier suppliers in many different locations: a practice referred to as "follow sourcing." Today, the criteria used by the OEMs to assess whether a supplier will retain Tier 1 status focus on price (both initial price and the ability to achieve successive reductions in price over the life of the contract), quality, ability to deliver consistently, technological capability, geographical reach of the supplier (i.e., the supplier's ability to follow source), and the supplier's ability to manage and coordinate lower levels in the supply chain.

As noted above, the number of direct first-tier suppliers has shrunk considerably. The last decade has seen some major suppliers who formerly supplied directly to the OEM either retreat to become second-tier suppliers or exit the industry entirely. More recently, first-tier suppliers have begun aggressively to cull their own lower-tier suppliers using criteria similar to those set by the OEMs in their winnowing of first-tier suppliers. One consequence has been an increase in merger and acquisition activity. Some first-tier suppliers have acquired second-tier firms that they see as being strategic for their continued development as first-tier suppliers of modular systems. Mergers and acquisitions have also occurred among second- and third-tier suppliers as firms at lower levels in the supply chain struggle to survive the restructuring under way in the industry.

The tight time management of the supply chain demanded by modular assembly has forced many first-tier suppliers to establish new production facilities in close proximity to OEM customers who require modules to be supplied in a sequence closely synchronized to the production process in their assembly plant.[11] Many of the plants owned by companies such as Lear, Dana, TRW, Meritor, and Johnson Controls that make up the foreign-owned component manufacturers segment of the automotive parts industry in Canada are of this kind. The future size of this segment of the industry depends largely on the extent to which Canada remains an attractive place to invest and on Canada's ability to attract new assembly investment (Pilorusso 2002).

There is a small group of Canadian-owned automotive parts manufacturers that grew rapidly over the last two decades to become important North American suppliers. First and foremost is the Magna family of companies — Magna International, Intier, Decoma, Tesma, and Magna Steyr — that are truly globally competitive first-tier suppliers.[12] Other Canadian companies which rank as major suppliers include Linamar, Wescast, Meridian, the ABC Group, the Woodbridge Group, and AG Simpson. However, a large proportion of Canadian-owned parts manufacturers are second-tier suppliers that to date have had limited geographical reach. While these firms may have done well in the past, many now face significant challenges as assembler-supplier relationships are restructured and as each tier is squeezed by the next highest level in the automotive food chain.

Innovation in the Automotive Parts Industry

Since the focus of the broader ISRN project is on regional systems of innovation, what can be said at this stage in our research regarding the general nature of innovation and research and development in the automotive parts industry? Within the global automotive industry and the broad North American auto industry, significant ongoing effort and resources are being directed to both product and process innovation. However, commentators repeatedly note that research and development (R&D) spending in the Canadian auto industry is meager at best and this is borne out by available data.[13] In large measure, this is a reflection of the structure and historical evolution of the North American auto industry. The Canadian and US industries became fully integrated following the creation of the Auto Pact in 1965.[14] As a result of integration, most of the research, development and engineering expertise within the Big Three automakers that existed in Canada prior to the Auto Pact became centralized in the United States, particularly in Michigan and California. As noted earlier, until the last decade or so in addition to overall vehicle design the assemblers also undertook most of the design, development, and testing of parts. The suppliers only carried out the manufacture of parts to the designs and specifications of the assemblers. Today, first-tier suppliers are expected to participate in vehicle platform development teams from project initiation. Desrosiers stresses that "in-house design, development and testing facilities, and a sales office and/or R&D centre in Detroit are absolute requirements for parts companies wishing to function as system integrators and to be included on the teams" (n.d., p. 34).

In a review of general lessons emerging from the numerous case studies under the umbrella of the ISRN project, Wolfe and Gertler (2003) identify two distinctive "models" of cluster development into which the case studies seem to fall: the regionally embedded and anchored cluster, and the entrepôt type cluster. In the former,

> while global knowledge flows are certainly important to the competitive success of local firms, the local knowledge/science base represents a major generator of new, unique knowledge assets. Local universities and research institutes constitute an important part of this base as "anchors" that generate highly skilled graduates, spin-off start-ups, and new, publicly available knowledge ... In many cases, there appears to be one or a few "anchor" firms or lead institutions that play a critical role in these processes (Wolfe and Gertler 2003, p. 28).

In the case of the entrepôt cluster, much of the knowledge base required for innovation and production is simply acquired through market transactions, often from non-local and even global sources. Even so, local institutions and actors still play a crucial role in enabling firms located within the local cluster "to exploit this knowledge effectively and combine it with other local assets and capabilities for success" (ibid., p. 29).

From the limited research we have completed to date, it would appear that the entrepôt model probably best characterizes what we would call the "core" automotive parts industry in southern Ontario. First-tier, and increasingly second-tier suppliers, must now engage in innovation and R&D activity. However, since virtually all of the first-tier supplier plants in Ontario are embedded in large transnational corporations, much of the formal R&D activity takes place elsewhere within the corporate structure, often in facilities located in Michigan or other parts of the United States. The interviews that we have completed with "core" auto parts plants support this view. While many plant managers refer to incremental innovation, particularly incremental process innovation, occurring within their plant, most formal R&D and especially product development is carried out elsewhere within the company.

The relative importance of incremental innovation rather than formal R&D means that the southern Ontario automotive parts firms rely heavily on their workforce for product and process improvement. However, available evidence suggests that during the 1990s, outside of a handful of large and mostly unionized component manufacturers, investment in training and other human resource strategies remained a relatively low priority for automotive parts-makers (see Canada Consulting Cresap 1991; Paget Consulting Group 1996). As such, automotive parts producers have relied on the public secondary and postsecondary education system for skill development. Studies show that Ontario workers have a higher level of educational attainment than in the US (Ontario. MEDT 2001). Indeed, in 1997 Canada ranked third after Germany and France with regard to the availability of skilled workers and considerably ahead of the United States which ranked seventh (Charles River Associates 2001). Furthermore, employee turnover, which research suggests is conducive to firm learning (Tomlinson 2002), is lower in Ontario than in the US. This may be associated with the higher level of unionization in the Canadian auto parts industry, which is almost double that in the United States (45–50 percent versus 20–23 percent). The overall impact of higher levels of unionization in the Canadian automotive parts industry on innovation and competitiveness is unclear. However, our interviews suggest that, for at least some firms, unions provide an important conduit of knowledge between different workplaces; knowledge that plant managers view as a source of innovative practice.

In an effort to foster collaborative R&D with the Big Three and establish Windsor as the "Automotive Intellectual Capital of Canada," the University of Windsor has established several research centres and Industrial Research Chairs linked to automotive engineering over the last ten years. However, the University's collaborative research efforts are largely geared to the OEMs; the plant managers we have interviewed in both the core automotive parts sector and the MTDM industry in Windsor have made only limited reference to the role played by the University of Windsor in the automotive parts cluster. In contrast, they have frequently stressed the importance to their companies of the pool of well-trained skilled production workers produced by St. Clair Community College and, in the past, area technical high schools.

The mould-making sector in the Windsor-Essex County region is very different and in many respects resembles a classic, regionally embedded and anchored cluster. The industry has developed over the last 40 years and now consists of a dense network of over 150 small and medium-sized firms many of which were spun-off from a large anchor firm that no longer exists. Many of the firms are highly specialized, privately owned by entrepreneurs with strong skilled technical backgrounds, and employ highly skilled mouldmakers. While there is little formal cooperation or information sharing among these highly competitive firms there nevertheless appears to be a significant amount of informal knowledge that flows between firms through the movement of skilled workers between plants and through social and family networks. Most innovation appears to be incremental and to originate from the skilled workers on the shop-floor rather than through more formal R&D activities. Although initially built by and with skilled immigrant workers, the industry now relies heavily on local secondary educational institutions and the community college to provide the necessary supply of skilled labour.[15]

RESTRUCTURING AND PERFORMANCE OF THE CANADIAN AUTOMOTIVE PARTS INDUSTRY IN THE 1980S AND 1990S

Virtually all of the key economic indicators reflect the significant expansion and impressive performance of Canada's automotive parts industry during the 1980s and 1990s. Besides the strong growth in the value of shipments, labour productivity, and employment noted earlier, both imports and exports increased in volume during the 1990s.[16] Perhaps significantly, the only indicator that barely increased over the 20-year period was real wages.

This expansion was driven by several factors. First, assembly plants in Canada fared well with regard to new investment and expanded production of popular models that sold well in the booming US market. The growth in assembly output, together with the increasing pressure on key first-tier suppliers to locate in close geographical proximity to their assembly plant customers, produced a trickle-down effect to suppliers. However, two-thirds of Canadian-made parts are shipped to the US, so the growth in the parts industry cannot be explained solely by growth in the Canadian assembly sector. A second significant factor in the growth was the decline in the value of the Canadian dollar, making exports of Canadian-produced components cheaper for US customers. Based on our own research, empirical evidence suggests that a third major factor was the improved competitive position of Canadian-based producers. The latter resulted from the far-reaching restructuring of the Canadian automotive parts industry that commenced in the 1980s and gained momentum in the 1990s in response to mounting competitive pressures. This restructuring of the industry involved less efficient plants exiting the industry, investment in new manufacturing capacity, and the modernization of many existing plants. The restructuring resulted in a more efficient, technologically sophisticated industry that was more consolidated and exhibited a significantly higher level of Canadian ownership. However, the application and the outcomes of these industry-wide structural changes were far from uniform; the diversity that exists among the component sub-industries and the historical peculiarities of the Canadian industry mean that the complete picture is much more complex.[17]

As part of another project, we analyzed the dynamics of restructuring and industrial competition in the automotive parts industry in the 1980s and 1990s and their impact on the structure and composition of the industry.[18] Briefly, the conceptual framework and datafile used for that analysis are as follows. We have already noted that plants in the competitive automotive parts industry exhibit significant heterogeneity with regard to size, age, technology, and productivity. They also experience high rates of turnover (establishment "births" and "deaths"). Models that incorporate the Schumpeterian notion of "creative destruction" account for such phenomena by linking firm dynamics, competition, economic growth and change in an industry. In a competitive industry, such as automotive parts, new manufacturing establishments continually enter the industry and some existing companies close or switch their production to another industry. Successful entrants not only introduce new or advanced technology but also exert competitive pressure on incumbent establishments to become innovative. Competition progressively weeds out the less successful organizations and favours those newly entered establishments capable of competing with the incumbents. Thus, new entrants tend to drive innovation forward in an industry, which, in turn, leads to aggregate productivity increases as the process of entries and exits "reallocates resources from losers to winners."[19] Over a period of years, the cumulative impact of entries and exits can bring about a profound shift in the characteristics and performance of the establishments that make up an industry.

Our analysis of the dynamics of industrial competition in the automotive parts industry used a Statistics Canada longitudinal file that allows the analysis and comparison of the characteristics of plants entering and leaving the industry as well as continuing or incumbent plants.[20] Records on firm ownership and other variables such as technological diffusion can be linked to the longitudinal datafile. Here, we briefly summarize findings from that project that provide useful context for our ISRN study.

New Investment and Modernization

Through the 1980s and 1990s, massive capital investment and the introduction of new production technology took place in both new and existing auto parts plants.[21] The 1998 Advanced Manufacturing Technology Diffusion Survey conducted by Statistics Canada revealed a significant take-up of new technology within the automotive parts industry. By linking the Statistics Canada longitudinal datafile with the 1998 Advanced Manufacturing Technology Diffusion Survey, we have been able to compare the adoption of new technology by plant age, plant size, firm structure, location, and ownership. In general, plants established in the 1990s were more likely to be using new technologies than older plants, as were large plants compared to small plants. Plants owned by multi-plant enterprises were more likely to be using new technologies than single-plant firms. Plants using new technology were more often located on the edge of metropolitan areas. However, the degree of technology diffusion was similar between Canadian- and foreign-owned plants.

TABLE 2
Shares of Employment, Shipments and Value-Added by Establishment Size, Automotive Parts Industry, Canada, 1999

Size of Establishment by Employment	*Number of Establishments (% of total)*	*Production Employees (% of total)*	*Value of Mfg. Shipments ($000,000) (% of total)*	*Mfg. Value-Added ($000,000) (% of total)*
1–49	243 (44.7)	3,384 (4.1)	492,875 (1.7)	232,086 (1.9)
50–99	61 (11.2)	4,034 (4.9)	766,660 (2.7)	355,165 (3.0)
100–199	89 (16.4)	11,715 (14.1)	2,668,550 (9.3)	1,208,215 (10.1)
200+	151 (27.8)	63,700 (76.9)	24,633,270 (86.2)	10,134,741 (85.0)
Total	**544**	**82,833**	**28,581,355**	**11,930,207**

Notes:
1. This table excludes motor vehicle plastic parts.
2. The year 1999 is the most recent year for which Statistics Canada data by size of establishment are available. Also, in 2000, Statistics Canada incorporated major conceptual and methodological changes into the *Annual Survey of Manufactures*, which led to the inclusion of a large number of small establishments not previously included in the ASM. The total number of establishments in NAICS 3363 increased from 544 in 1999 to 884 in 2000 largely as a result of this change in data collection.
Source: Statistics Canada, special tabulation by authors.

Labour Costs, Productivity and Competitiveness

Much has been made of the competitive threat to the Canadian auto parts industry posed by plants in the United States and Mexico. While Mexican plants enjoy a narrow cost advantage on parts with a very high labour content, the real competitive threat for the industry in Canada is the United States. The growth in labour productivity in the parts industry as a whole in Canada kept pace with that in the United States through the 1990s, but the gap in labour productivity between the United States and Canada persisted. The factors underlying this productivity gap need to be better understood. For example, given that there is significant variability in labour productivity between sub-industries and between plants at different tiers in the supply chain, to what extent is the "productivity gap" between Canada and the US a reflection of a different mix in sub-industries and/or first- versus second-tier plants between the two countries? That the Canadian industry grew despite the productivity gap, again raises the question of the impact that the currency exchange rate has on labour costs and competitiveness. Expressed in national currencies, the average hourly wage in the automotive parts industry rose faster in Canada than in the United States although real hourly wages in both countries remained relatively flat. Due to the falling value of the Canadian dollar during the 1990s, when average real hourly wages are converted to US dollars, Canada continued to enjoy a labour cost advantage of approximately 20–25 percent.[22] Does the labour cost advantage cushion competitive pressure on Canadian plants and hence result in lower levels of labour productivity?[23]

Domestically, a labour productivity gap also exists within the Canadian auto parts industry between Canadian-owned and foreign-owned plants. On the surface, Canadian-owned plants have done well, increasing their share of number of plants, employment, and value of shipments. Much of the growth in employment that occurred in the 1990s came from Canadian-owned plants (Figure 2). However, Canadian-owned plants continue to lag behind foreign-owned plants in both labour productivity and wages. On average, foreign-owned plants in Canada have larger workforces, are more likely to belong to a multi-plant firm, have higher wage rates and levels of labour productivity, and lower wage-to-total cost ratios than Canadian-owned plants.

Finally, we should underscore the heterogeneity that exists between sub-industries, across a number of measures within Canada. Over 50 percent of all Canadian plants make plastic parts, engines and engine parts, or metal stampings. The strongest employment growth occurred in the metal stamping, plastic parts, and seating/interior trim sub-industries, whereas the largest increases in labour productivity took place in the engines/engine parts, seating/interior trim, and transmission and power train sub-industries. Using value of shipments as an indicator, the largest increases in value of shipments occurred in the engines/engine parts, seating/interior trim, transmission and power train, and plastic parts sub-industries (Fitzgibbon and Holmes 2003). Examining auto parts industry productivity and competitiveness by sub-industry is more likely to yield an accurate picture than looking at the industry as a whole.

FIGURE 2
Total Employment, Canadian-owned and Foreign-owned Plants, Automotive Parts Industry, Canada, 1980–99

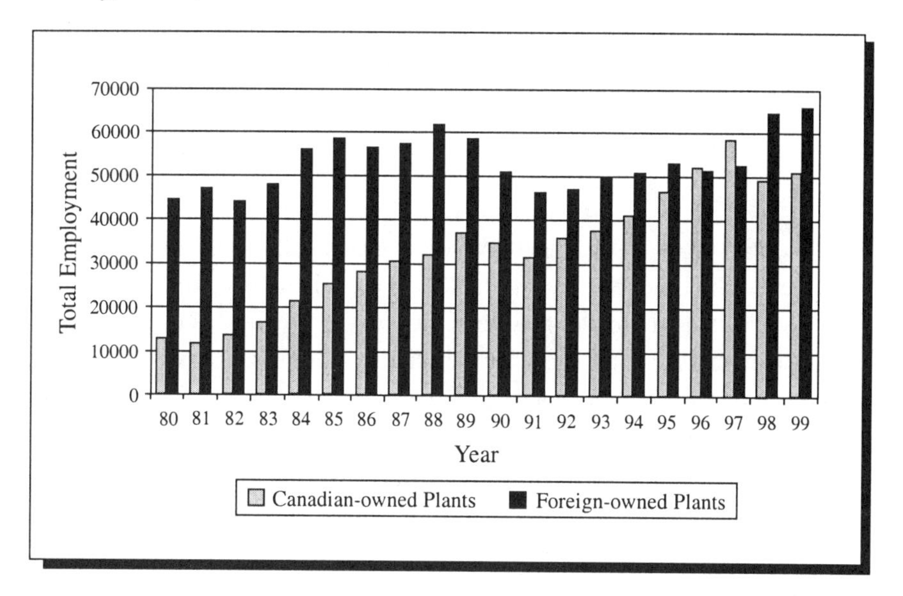

Note: This is based on SIC 325.
Source: Statistics Canada, special tabulation by authors.

CHALLENGES FACING THE AUTOMOTIVE PARTS INDUSTRY IN ONTARIO

Output and employment in Canada's automotive parts industry peaked in 2000 and then levelled off between 2000 and 2002. The parts industry, however, has not experienced nearly as severe a downturn as the vehicle-assembly industry in Canada.[24] Nevertheless, concerns have resurfaced regarding the future prosperity of the industry in Ontario and many commentators view the industry as being at a critical crossroads. These concerns stem from the conjuncture of several recent developments which have significantly increased competitive pressures and uncertainty in the industry. These developments include the final demise of the Auto Pact, the continuing decline in the Big Three automakers' share of the North American motor vehicle market, the relative shift during the 1990s of vehicle-assembly capacity from the Great Lakes region toward the

US south and Mexico, the significant increase in the value of the Canadian dollar against the US dollar since mid-2002, and, especially, the change in assembler-supplier relationships and supply-chain management practices discussed earlier in this chapter. Given the impending wave of retirements among skilled workers, there are also crucial skills and training issues looming, especially in the more skilled areas of the industry such as tool, die, and mould.

The End of the Auto Pact

The Auto Pact triggered a wave of new investment in both the assembly and parts industries in Canada during the late 1960s and early 1970s and remained the cornerstone for automotive policy and the further expansion of automotive production in Canada during the 1980s and 1990s (Holmes 1983, 1992, 2004*b*). In February 2001, the World Trade Organization (WTO) struck down the Auto Pact, ruling that it violated international trading rules (Anastakis 2001). What consequence did the end of the Auto Pact have for the automotive parts industry in Canada? Under the Auto Pact, motor vehicles and original equipment parts entering Canada from the United States or other countries did so duty free as long as the importer met the performance requirements necessary to be classed as a "qualifying Canadian manufacturer." These included both minimum production volumes for assembled vehicles and certain minimum levels of Canadian value-added (CVA) content within the vehicles produced. The CVA requirement was designed to ensure that Canadian production amounted to more than mere assembly, and helped generate demand from the Big Three for automotive parts made in Canada.[25] Since the demise of the Auto Pact there is no longer any special incentive for automakers to purchase Canadian-made parts and parts-makers in Canada are subject to the full force of international competition.

The Big Three's Shrinking Share of the North American Vehicle Market

The automotive parts industry in Canada has always been heavily dependent on the Big Three for its customers. Today, the Big Three still account for over 75 percent of vehicle-assembly capacity in Canada and over 90 percent of Canadian parts purchases (CAW 2003). The arrival in North America of the Japanese-based automakers in the 1980s and 1990s saw relatively few Japanese suppliers open new plants in Canada compared to the United States and existing Canadian parts plants had difficulty winning contracts from the Japanese OEMs. Over the last decade, the share of the North American vehicle market accounted for by the Big Three has declined steadily with the decline accelerating in recent years, falling from 70 percent in 1999 to 62 percent in 2003 (*Automotive News* 2003). Not surprisingly, the Big Three's share of total production also declined as more new assembly plants have been built in North America by Asian- and European-based automakers. By 2003, over 28 percent of all the vehicles produced in North America were assembled in plants belonging to non-Big Three producers (*Automotive News*, 5 January 2004, p. 42). In other words, the Big Three market upon which Canadian automotive parts companies have traditionally relied has been steadily eroded.

The Shifting Geography of North American Vehicle-Assembly Capacity

Linked to this market shift from the Big Three to Asian- and European-based automakers has been a geographical southward shift in the centre of gravity of North American assembly capacity. Mexico's auto industry grew rapidly in the 1990s, competing with Canada for new investment. At the same time, states such as Alabama, Mississippi, and South Carolina began to offer unprecedented investment incentives to lure new assembly plants, particularly those of Asian and European OEMs, to feed their growing shares of the North American market (Klier 2003). Of the 18 new assembly plants built or announced in North America since 1990, only three are in the Great Lakes region and only one of those is located in Canada. At the same time, existing Big Three assembly plants in the Great Lakes region have closed.[26] We noted earlier that there is considerable pressure on first-tier suppliers of major systems and modules to locate in close geographical proximity to their customer's assembly plants. Therefore, in response to the shift in assembly capacity toward the American south and Mexico, many first-tier suppliers are forced to follow suit if they are to win the contracts to supply modules to the new plants. Thus, automotive parts manufacturers in Ontario are not only seeing their traditional Big Three market shrink, but are faced with the challenge of increasing their sales to Asian- and European-based OEMs whose production capacity is increasingly located at a considerable distance from Ontario.

During the expansion of the automotive industry in Canada during the 1980s and 1990s, Canada enjoyed a distinct production-cost advantage due to a combination of the low value of the Canadian dollar, the public health-care system which, in comparison to the United States, saves automakers up to $6 per hour worked, and high productivity linked to the more highly skilled labour force. The more than 20 percent appreciation in the value of the Canadian dollar against the US dollar over the last 18 months, rising from US$0.63 to US$0.78, has eroded the cost advantage and begun to put enormous competitive pressure on Ontario parts-makers.[27]

Squeezing of Suppliers

Perhaps the greatest source of uncertainty with regard to the future vitality of the industry has been generated by the OEM's focus on cost-cutting across the supply chain and the associated changes in assembler-supplier relationships we discussed earlier. These have left many parts companies scrambling to redefine their role in a rapidly changing industry. In virtually all of the interviews we have conducted to date with plant mangers, supply-chain issues have been the greatest concern.

The Big Three OEMs are becoming more aggressive in demanding annual price cuts from their suppliers.

> Ford Motor Co. and General Motors Corp., both under pressure to boost profits, are setting rigorous new targets for price cuts by their parts suppliers. The moves could force parts makers in developed countries to cut their prices to close the gap with rivals in low-wage countries like China ... as a first step, Ford said it expects price cuts of 3.5 per cent from its suppliers by Jan. 1 [2004], with more to follow ... GM, meanwhile, has warned suppliers that in order to retain business with the Detroit automaker, a supplier would have 30 days to come up with corrective measures if its price is found to be higher than that of a rival's (Shirouzu 2003, p. B16).

Suppliers are expected to achieve cost-savings by re-engineering components, finding cheaper material, or coming up with simpler designs. This is putting considerable pressure on the domestically owned segment of the Canadian industry, dominated as it is by second-tier suppliers which currently have little in-house research and design capability.[28]

Similar pressures are also being experienced by the suppliers of tooling to the automotive industry. Mouldmakers interviewed in Windsor are concerned about cash-flow problems and issues related to intellectual property rights resulting from the changing practices of OEMs and first-tier suppliers. The design, engineering, testing, and production of some moulds can take more than a year to complete and cost upward of $1 million. Historically, the mouldmaker would receive part payments from the customer at a number of different times during production. Now, the mouldmaker is often forced to wait until the mould has actually gone into production at the customer's plant before receiving any payment. Since the vast majority of mouldmakers in Windsor are small to medium-sized enterprises with limited external sources of financing, this change has created serious cash-flow problems. The Windsor mouldmakers are highly skilled, innovative, and regarded as North American leaders in their industry. Until recently, however, there was limited formal patenting activity since the ownership of intellectual property was not a major concern and the patenting process was expensive. Mouldmakers are now concerned that customers are increasingly using their expertise in Windsor to design and engineer the mould and then giving that intellectual property to companies in Asia, and especially China, to produce the tooling at lower cost. Mouldmakers point out that should this trend continue, it will significantly undermine the innovative capacity of the industry in Windsor since so much of the expertise and innovation in mouldmaking arises from the hands-on experience of highly skilled production workers.

Supply and Retention of Skilled Workers

Another major issue for both the automotive tooling industry and the larger, more innovative parts-makers is the supply and retention of skilled workers. A recurring theme in many of the interviews we have conducted in Windsor is the poaching of workers in the skilled trades. After being trained in the tooling and mould shops and the auto parts plants, workers are either lured away by the prospect of higher wages and greater job security offered by the Big Three plants in Windsor or by higher paying jobs across the border in Michigan to which they can commute. The demography of the skilled labour force is also an issue, especially for the automotive tooling industry. Many of the entrepreneurial, skilled workers, who founded and built the highly successful mould-making industry in Windsor around privately owned, internally financed, small companies, are nearing retirement and very little succession planning has taken place to ensure that these companies will continue to flourish and prosper once their current owners have retired.

CONCLUSION

We have endeavoured to provide an overview analysis of recent developments in, and the challenges that currently confront, the automotive parts industry in Ontario. The auto industry is not only one of the most important and key manufacturing industries in Canada but its geographical concentration in southern Ontario makes it a mainstay of that province's economy. The enormous expansion of both vehicle-assembly and automotive parts manufacturing in the 1990s underpinned much of the prosperity enjoyed by southern Ontario during that period. However, a number of recent developments have called the continued future vitality of the industry into question.

A recent CAW pamphlet succinctly sums up the experience of the industry over the last decade:

> The independent auto parts sector typifies the best, and the worst, that capitalism offers working people. On the positive side, the Canadian parts industry has been a beehive of innovation, technological advances, and entrepreneurship over the last decade. Canadian-based firms have expanded rapidly, both in the quantity of output and employment, and in the quality and sophistication of the products the industry makes .... on the negative side, however, the auto parts industry also reflects the never-ending battle for survival that is an inherent feature of a competitive, capitalist economy. Even when the parts industry was booming in Canada in the 1990s, workers in many companies never felt secure, thanks to the ongoing restructuring of the industry, vicious competition between parts companies, and relentless pressures of technical change (CAW 2003).

We have shown that the dynamics of industrial competition, as reflected in high plant entry and exit rates, significantly changed the composition and improved the economic performance of the parts industry in Canada during the 1980s and 1990s. Consolidation led to an increase in the average size of plants, and output and employment became increasingly concentrated in larger plants. The relative proportion of industry employment and output accounted for by Canadian-owned plants increased sharply as a result of the aggressive growth and expansion of a handful of companies, most notably the Magna family of companies, which emerged as globally competitive suppliers to the world auto industry.

Besides the restructuring, much of which was driven by changes in assembler-supplier relationships introduced by the OEMs, the outstanding performance of the parts industry in Ontario in the 1990s was also underpinned by the growth in vehicle assembly in Ontario and the province's competitive cost position within the North American auto industry. However, over the last three years several clouds have appeared on the horizon. The rising value of the Canadian dollar, the Big Three's continued loss of market share, the southward shift in vehicle-assembly capacity, the downloading from OEMs to suppliers of R&D, design, and supply-chain management functions, and the continual price squeezing by higher tiers in the supply chain have all conspired to introduce uncertainty with regard to the future of the industry.

Both the APMA and the CAW have called for governments to develop an economic strategy to protect the jobs that were created in the parts industry during the 1990s and to enhance the longer-term prospects for parts-makers located in Canada. Stemming the rising value of the dollar, skills and training, and reducing the delays in the flow of automotive products across the Canada-US border have been identified as important areas in which government action is required. Many Canadian parts firms do not currently have either the technical or financial capability to undertake the increased levels of R&D being demanded by the OEMs. This is another key area in which governments could provide assistance.

Despite being a mature industry, the automotive parts industry in Ontario has experienced very significant technological changes, innovation, and entrepreneurship over the last decade. In subsequent papers drawn from our study we will provide a more in-depth analysis of the institutional mechanisms that have fostered innovation and growth in the parts industry in two local economies in southern Ontario — Windsor-Essex County and Kitchener-Waterloo-Cambridge-Guelph — that rely heavily on the automotive industry.

NOTES

[1]Some 35,000 new jobs were created in the parts industry between 1992 and 1999 alone (CAW 2003).

[2]The Institute for Competitiveness and Prosperity analysis defined the "automotive cluster" as consisting of seven narrow sub-clusters (motor vehicles, automotive

parts, automotive components, transportation equipment, flat glass, forgings and stampings, and production equipment) and six broad (metal processing, related machinery, related vehicles, other engines, related parts, and related equipment) covering a total 32 industries. As we shall see shortly, defining the automotive cluster, either industrially or geographically, is not an easy or straight-forward exercise.

[3]These actors include representatives of the automakers (the customers of the automotive parts industry) involved in supplier base development, the key manufacturers business association (the Automotive Parts Manufacturers Association [APMA]); the two unions that represent the vast majority of unionized workers in the industry (the Canadian Autoworkers [CAW] and United Steelworkers of America [USWA]), provincial and federal government officials, and private-sector consultants who specialize in the automotive industry.

[4]The corresponding codes under the previous SIC classification system used by Statistics Canada were SIC 3231 (motor vehicle assembly) and SIC 325 (automotive parts and accessories). One important difference between the SIC and NAICS classification systems is that the automotive plastic parts industry was included in SIC 325 but is not included under NAICS 3363. With respect to the auto parts industry as a whole, there is a relatively high concordance (on the order of 90–95 percent) between the old SIC system and NAICS, especially if the motor vehicle plastic parts industry (NAICS 326193) is combined with NAICS 3363.

[5]For the purposes of our analysis, unless otherwise stated when we present Statistics Canada data pertaining to the automotive parts industry, the data are combined figures from the eight sub-industries within NAICS 3363 plus motor vehicle plastic parts (NAICS 326193).

[6]It is estimated that 80 percent of Canadian mouldmakers are located in Ontario with 50 percent of those located in Windsor (USITC 2002, p. 4-3). From our interviews with mouldmakers in Windsor it appears that at least 80 percent of their customers are in the automotive industry. About 80 percent of the Canadian die sector is also located in southern Ontario (ibid.). We estimate that there are approximately 250 tool, die and mould establishments in the Windsor area (authors' database) the vast majority of which are linked to the auto industry. Desrosiers (n.d., p. 82) reported that employment in the Windsor-Essex County machine, tool, die, and mould sector more than doubled between 1991 and 1999, increasing from 4,400 to 9,140. In the Windsor local economy there are also a number of other firms that supply various business services to the automotive industry. This will be explored further in our more detailed study of the "automotive cluster" in Windsor.

[7]In 2001, the most recent year for which data are available, 90 percent of the employment and 95 percent of manufacturing shipments in the Canadian automotive parts industry were found in Ontario. Although Quebec had 18 percent of the plants in the industry it accounts for less than 6 percent of employment and less than 3 percent of manufacturing shipments (CANSIM Table 301-0003).

[8]Desrosiers states that the growing MTDM sector in Windsor "is not only a significant generator of employment in its own right, but also a powerful draw to

suppliers of automotive aluminum, iron and magnesium castings, moulded plastic components and stampings. This has long been recognized by the Windsor-Essex County Development Commission, which has focused on MTDM as a cornerstone of its automotive business development strategy" (n.d., p. v). The Windsor mould-making sector, which is an acknowledged technological leader, exports much of its production outside the region and, hence, does not see its future to be linked so directly to the health of the local assembly sector as does the "core" automotive parts industry.

[9]In 1999, assembly in Canada peaked at over 3 million vehicles. This ranked Ontario as the second largest automobile producing jurisdiction in North America behind Michigan, and Canada as the fourth largest auto-making country in the world (MEOI 2003). Since then, assembly has declined by 15 percent and over 7,000 jobs have been lost from the assembly industry (CAW 2003).

[10]This trend involves the monitoring of large volumes of data pertaining to production and quality standards increasingly collected over the Internet. Specialized e-business Internet portals such as COVISINT (www.covisint.com) have emerged to assist OEMs and large first-tier suppliers to manage and monitor the supply chain.

[11]This is particularly true for the production of those modules that must be delivered in strict colour sequence to the OEM; for example, seating modules, interiors, instrument panels, fascias, etc.

[12]Magna International ranks as the fifth largest parts manufacturer in North America by OEM sales (*Automotive News* 2003) and is far larger than the next largest Canadian-owned parts-makers Woodbridge Group and Linamar which rank 54th and 56th respectively. Magna's sales are ten times greater than those of either Woodbridge or Linamar.

[13]In a recently published list of Canada's top 100 corporate R&D spenders there was only one automotive parts firm, Magna International. Although Magna ranked second in absolute number of dollars spent on R&D in 2002 this represented only 2.8 percent of revenue (Research Infosource 2003).

[14]The Canada-United States Automotive Products Trade Agreement, commonly referred to as the Auto Pact, was a managed free trade agreement which allowed duty-free movement of assembled vehicles and automotive parts between Canada and the United States whilst ensuring certain continued levels of automotive production would remain within Canada (Holmes 1983, 2004*b*; Anastakis 2000).

[15]The Windsor tool, die, and mould-making industry will be analyzed in more depth in a future paper once the interviewing phase of our study is complete.

[16]The increase in both exports and imports is a reflection of the high level of production integration within the North American auto industry. Measures of export intensity (exports/shipments) and import penetration (imports/estimated domestic market — shipments+imports–exports) remain high. They did decline, however, during the 1990s suggesting that a slightly greater proportion of auto parts produced in Canada went into domestic vehicle assembly by the end of the decade. There was significant growth in the size of the markets for the engines/engine parts, transmission and drive train, and seating/interior trim sub-industries, directly reflecting the expansion of vehicle assembly in Canada.

[17]While space precludes further elaboration of this point, variations between sub-industries are discussed in detail elsewhere (Fitzgibbon and Holmes 2003).

[18]This project is titled "The Dynamics of Restructuring and Industrial Competition in the Canadian Automotive Parts Industry" (J. Holmes and P. Kumar, co-PIs) and is funded by the AUTO21 Network of Centres of Excellence and the Center for Automotive Materials and Manufacturing (CAMM).

[19]This evolutionary process not only affects the structure and performance of establishments and firms, but also impacts wages, working conditions, work organization, work content, the nature of jobs, and the industrial relations climate.

[20]The datafile contains longitudinal data for all Canadian manufacturing plants during the period 1973–99. The file is derived from the *Annual Survey of Manufactures* that is sent to all large plants and from administrative tax data for small plants. It contains establishment (plant)-level data on employment (production and non-production), wages and salaries, hours paid, manufacturing and total shipments, manufacturing and total value-added, and components of total production costs. For a detailed description of the datafile and its utility for analyzing the dynamics of industrial competition see Baldwin (1995). Fitzgibbon, Holmes and Kumar (2003) present preliminary results from our analysis of the competitive dynamics within the automotive parts industry. We acknowledge the valuable assistance with the data analysis provided by Dr. John Baldwin, Director, Microeconomic Analysis Division, Statistics Canada.

[21]Total new capital investment in Canada's automotive parts industry between 1991 and 2001 amounted to over $10.34 billion dollars (Statistics Canada 2004*b*).

[22]Just as there is significant variability in labour productivity between sub-industries there is also significant variation in hourly wage rates. So, once again, the difference in average hourly wage rates between the industries in Canada and the United States may in part result from a different mix of sub-industries between the two countries.

[23]Note that the possible strategies for enhancing productivity and competitiveness are fundamentally different depending on the balance between the different possible explanations for the productivity gap outlined in this paragraph. One could seek ways of boosting productivity in general or, if the gap is due to structural differences between the US and Canada, the effort might be better directed at encouraging the growth of the sub-industries with higher labour productivity and higher value-added.

[24]Output in Canada's vehicle-assembly sector peaked in 1999 at over three million vehicles. Since then it has declined by over 15 percent with the closing of two assembly plants and the loss of over 7,000 jobs (CAW 2003). Recently, Statistics Canada reported that employment in the parts sector continued to recover from the dip in 2001 increasing by 6,872 jobs, or 6.9 percent, in the first six months of 2003 as compared with the same period in the previous year (van Alphen 2003).

[25]Although the 1989 Canada-United States Free Trade Agreement and the subsequent North America Free Trade Agreement (NAFTA) contained only a North

American value-added requirement, there remained an incentive throughout the 1990s for the Big Three to continue to meet the CVA requirements of the Auto Pact. The competitive advantage for the Big Three and CAMI of maintaining their Auto Pact status was that it allowed them to import vehicles and parts from third countries duty free (Holmes 2004*b*). During the 1980s, Canada granted duty-remission programs to a number of European and Asian automakers, which also boosted demand for Canadian-built auto parts. These programs were phased out under CUSFTA and NAFTA (ibid.).

[26]In Canada, while Honda and Toyota have expanded capacity at their Ontario plants, the Big Three's capacity has shrunk with the closing of plants by GM (Ste.Thérèse, QC) and Daimler Chrysler (Pillette Road Assembly, Windsor), and the planned closing in the summer of 2004 of Ford's Ontario Truck Plant (Oakville, ON).

[27]Many commentators and people we have interviewed view a US$0.70 exchange rate to be the critical level above which Canadian suppliers, and especially second- and third-tier suppliers producing commodity parts for which cost is crucial, come under severe pressure.

[28]Larger Canadian-owned and globally competitive first-tier suppliers such as the Magna family of companies — Linamar, the ABC Group, and Wescast — are perhaps better positioned to respond to this pressure but are not immune from it. Recently, Ford took the contract for front and rear fascias for the Freestar minivan built at Oakville Assembly away from Decoma and awarded it to a competitor, Flex-N-Gate, when Decoma refused to bow to Ford's demand to cut the price of the parts (Keenan 2003).

REFERENCES

Anastakis, D. 2000. "The Advent of an International Trade Agreement: The Auto Pact at GATT, 1964-1965," *International Journal* 55(4):583-602.

—— 2001. "Requiem for a Trade Agreement: The Auto Pact at the WTO 1999-2000," *Canadian Business Law Journal* 34(3):313-35.

Automotive Parts Manufacturers Association (APMA). 2002. *Major Assembler Investment Announcements*. Toronto: Automotive Parts Manufacturers Association.

Automotive News. 2003. *Automotive News Data Year Book 2003*. Detroit: Crain Publications.

Baldwin, J. 1995. *The Dynamics of Industrial Competition: a North American Perspective*. Cambridge: Cambridge University Press.

Canada Consulting Cresap. 1991. *Canadian Automotive Parts Industry Human Resource Study*. Ottawa: Employment and Immigration Canada.

Canadian Autoworkers Union (CAW). 2003. *Sectoral Profiles (Independent AutoParts; Major Auto)*. 7th Constitutional Convention, Toronto, August. Willowdale: Canadian Autoworkers Union.

Charles River Associates. 2001. *Competitiveness Factors for Attracting and Maintaining Automotive Investment: Comparison Between Canada and Mexico.*

Report prepared for Industry Canada and the Ontario Ministry of Economic Development and Trade, June.

Desrosiers. n.d. *Automotive Sector Strategic Plan: Research Report for Windsor-Essex County Development Commission*. Richmond Hill: Desrosiers Automotive Consultants Inc.

Fitzgibbon, S. and J. Holmes. 2003. "The Ontario Automotive Parts Industry." Paper presented at the Third Annual Conference of the Innovation Systems Research Network, Ottawa, April.

Fitzgibbon, S., J. Holmes and P. Kumar. 2003. "Changing Structure and Economic Performance of the Automotive Parts Industry in Canada." Paper presented at the Annual Conference of the Canadian Regional Science Association and the Canadian Association of Geographers, Victoria BC, May.

Flynn, M. and B. Belzowski. 1996. *The 21st Century Supply Chain: The Changing Roles, Responsibilities and Relationships in the Automotive Industry*. Chicago: A.T. Kearney Inc.

Holmes, J. 1983. "Industrial Reorganization, Capital Restructuring and Locational Change: An Analysis of the Canadian Automobile Industry in the 1960s," *Economic Geography* 59:51-71.

—— 1992. "The Continental Integration of the North American Auto Industry: From the Auto Pact to the FTA and Beyond," *Environment and Planning A* 24:95-119.

—— 2004*a*. "Re-scaling Collective Bargaining: Union Responses to Restructuring in the North American Auto Industry," *Geoforum* 35(1):9-21.

—— 2004*b*. "The Auto Pact from 1965 to the Canada-United States Free Trade Agreement (CUSFTA)," in *The Auto Pact: Investment, Labour and the WTO*, ed. M. Irish. The Hague: Kluwer Law International, pp. 3-21.

Institute for Competitiveness and Prosperity. 2002. "A View of Ontario: Ontario's Clusters of Innovation," Working Paper No. 1. Toronto: Institute for Competitiveness and Prosperity.

Keenan, G. 2003. "Ford Yanks Contract from Decoma: Dispute over Parts Prices Behind Move," *The Globe and Mail*, 20 December, p. B1.

Klier, T. 2003. "South Bound? Location Trends in the Auto Industry." Paper presented at the Auto21 conference "The Auto Industry in the 21st Century: Challenges and Prospects," Ottawa, April.

Ontario. Ministry of Economic Development and Trade (MEDT). 2001. *North America's Automotive Powerhouse*. Toronto: Ontario Investment Service, Government of Ontario.

—— Ministry of Enterprise, Opportunity and Innovation (MEOI) 2003. "2002 Was a Near-Record Year for Ontario's Auto Industry," *The Ontario Business Report* 7(4):1-16.

Paget Consulting Group. 1996. *Capitalizing on the Curriculum: The Viability and Long Term Plan for the Automotive Parts Sectoral Training Council*. Ottawa: Paget Consulting Group, F.J. Durdan and Associates.

Pilorusso, F. 2002. *Fostering Innovation and Investment in the Canadian Automotive Components Industry.* Research report prepared for Ontario Ministry of Economic Development and Trade and Industry Canada. Toronto: Pilorusso Consulting Inc.

Reguly, E. 2003. *Report on Business Magazine*, May.

Research Infosource. 2003. *Canada's Corporate Innovation Leaders.* At <http://www.researchinfosource.com>.

Shirouzu, N. 2003. "Ford, GM Push Suppliers for Aggressive Price Cuts," *The Globe and Mail*, 18 November, p. B16.

Statistics Canada. 2004*a*. *Annual Survey of Manufactures, Principal Statistics by North American Industry Classification System*, CANSIM Table 301-0003. Ottawa: Statistics Canada, 2 March. At <http://cansim2.statcan.ca>.

—— 2004*b*. *Capital and Repair Expenditures, Industry Sectors 31-33*, CANSIM Table 029-0009. Ottawa: Statistics Canada, 2 March. At <http://cansim2.statcan.ca>.

Tomlinson, M. 2002. "Measuring Competence and Knowledge Using Employee Surveys: Evidence Using the British Skills Survey of 1997," Discussion Paper No. 50. Manchester: Centre for Research on Innovation and Competition (CRIC), University of Manchester.

United States International Trade Commission (USITC). 2002. *Tools, Dies, and Industrial Molds: Competitive Conditions in the United States and Selected Foreign Markets*, USITC Publication No. 3556. Washington, DC: United States International Trade Commission.

Van Alphen, T. 2003. "Auto-parts Makers Add 6,872 Jobs," *The Toronto Star*, 12 September, p. E1.

Wolfe, D. and M. Gertler. 2003. "Lessons from the ISRN Study of Cluster Development," in *Clusters Old and New: The Transition to a Knowledge Economy in Canada's Regions*, ed. D.A. Wolfe. Kingston and Montreal: School of Policy Studies, Queen's University and McGill-Queen's University Press, pp. 1-36.

3

Learning, Innovation and Cluster Growth: A Study of Two Inherited Organizations in the Niagara Peninsula Wine Cluster

Lynn K. Mytelka and Haeli Goertzen

Introduction

Soil and climatic conditions have traditionally endowed specific regions with advantages in grape growing. As a result of the close linkage between grape growing and winemaking, wineries have tended to locate close to their suppliers. Over time ancillary services co-locate in the region forming a geographical agglomeration or cluster. In its emergence and development through the 1980s, the Niagara Peninsula wine cluster is typical of this model. Public policies contributed to shaping the cluster in this earlier period in two ways. On the input side, the Ontario Grape Growers Marketing Board (OGGMB), established in 1947 under the *Ontario Farm Products Act*, set minimum prices for and regulated the marketing of grapes grown in the province. On the output side, the Liquor Control Board of Ontario (LCBO), created in 1927 after Canada repealed prohibition, controlled the production and distribution of wine. The former took place through the Liquor Licensing Board of Ontario which licensed new wineries; the latter through the monopsonistic purchasing and sale of wine and alcoholic beverages in the province. These two organizations shaped the development of the Niagara wine cluster. This chapter examines their roles over this period through the lens of an innovation systems approach.[1]

FROM CLUSTER TO INNOVATION SYSTEM

A system of innovation is defined as a network of economic agents, together with the institutions and policies that influence their innovative behaviour and performance (Nelson 1993; Nelson and Winter 1982; Lundvall 1992). This approach characterizes innovation as an interactive process in which relations between firms and supporting organizations play a key role in bringing new products, new processes, and new forms of organization into economic use (see Figure 1). Following Edquist, Storper, and others, we understand institutions as "sets of common habits, routines, established practices, rules or laws that regulate the relations and interactions between individuals and groups" (Edquist 1977, p. 7), and as such they "prescribe behavioral roles, constrain activity and shape expectations" (Storper 1998, p. 24).

FIGURE 1
Innovation Systems

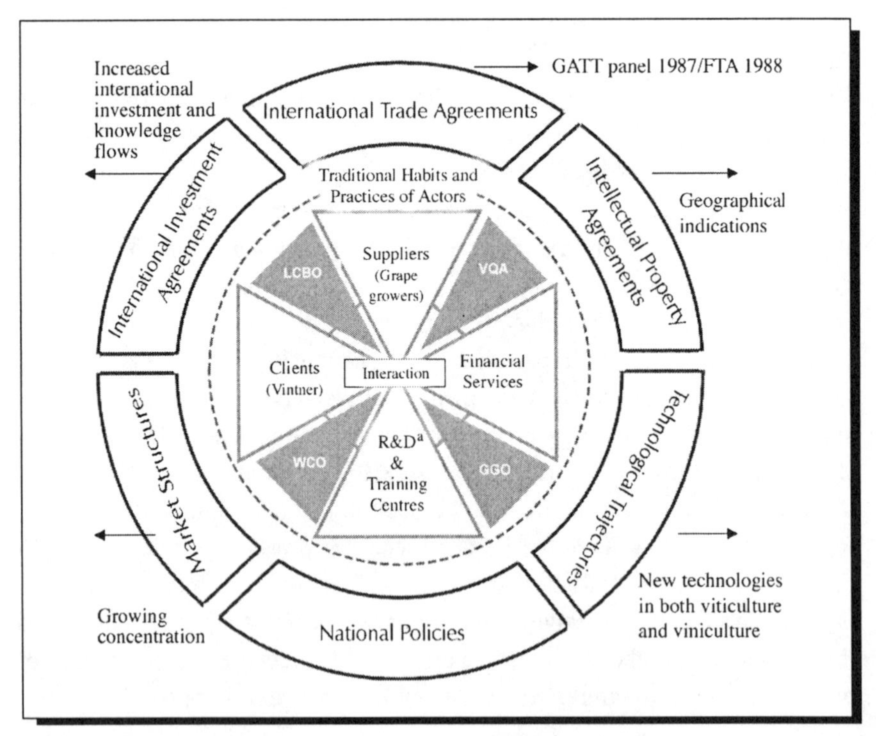

Note: [a]Brock University's CCOVI, Guelph University, Niagara College and Vineland Station.
Source: Mytelka (2000).

These habits and practices are learned behaviour patterns, marked by the historical specificities of a particular system and moment in time. As such, their relevance may diminish as conditions change. Learning and unlearning on the part of firms and other actors within the system are essential to the evolution of a system in response to new challenges.

The stimulus to innovation within a cluster might come from many sources. One is the market. Another is through non-market mediated collaborative relationships between individual users and producers of innovation (Lundvall 1988, p. 35) and where still broader systemic coordination is required, the stimulus to innovation might be intermediated by policies. Policies may stimulate and support a process of innovation by agenda-setting that targets specific actions for immediate attention. For example, between 1988 and 2000 the Ontario Wine Assistance Program (OWAP) radically altered the composition of grape varieties grown in the province by funding the removal of labrusca and hybrid vines and providing "sugar bonuses" to encourage the planting of vinifera vines and the use of newer viticulture techniques that produced high quality grapes. Since these transitional programs ended, demand for continued innovation in the grape sector has depended upon the relationships between vintners and grape growers. These relationships, too, are learned and the next section of this chapter analyzes the interactions within the Ontario Grape Growers Marketing Board, now the Grape Growers of Ontario (GGO), and cooperation within the cluster.

A variety of less-explicit policies also affect the innovation process by shaping the fiscal and legal parameters that constrain opportunities for new growth. The fourth section analyzes one of these parameters, the Liquor Control Board of Ontario's control over the sale and distribution of wine in the province. As a key actor within the emerging innovation system in the Niagara Peninsula, the LCBO has gained considerable prominence as one of the world's largest purchasers of wines and spirits. There is a long history in other innovation systems of the use of state purchasing power to stimulate innovation and support the growth of local enterprise.[2] Recent World Trade Organization (WTO) negotiations over subsidies now constrain this role to a large extent, but do not eliminate it altogether. Since the system's core actors operate under these parameters it is necessary to guage the impact that these policies have on innovation. Ontario will need to adapt local policies to this changing international environment. Adaptive policy-making is part of what makes an innovation system a learning system.

Within the above framework, it is possible to map the main actors in the Niagara wine cluster (Figure 1). Central to all innovation systems are supplier enterprises, here the grape growers, and their clients, the vintners. Other common actors within innovation systems are industry associations, research centres, standard-setting bodies, and education and training centres. All of these actors are present in the Niagara wine industry. Examples include the Wine Council of Ontario (WCO), the Cold Climate Oenology and Viticulture Institute (CCOVI) at Brock University and the Vineland Research Station now part of Guelph University,[3] the Vintners Quality Alliance (VQA), and Niagara College. Financial service providers such as banks and venture capital funds are also often an important part of an innovation system, but despite the significant investments needed to buy land, plant vineyards, and build wineries, they have not been important actors in the Niagara wine industry.

In mapping the Niagara Peninsula wine cluster, two organizations do not find their counterparts in either sector-based or cluster-based systems of innovation studied elsewhere. These are the Ontario Grape Growers Marketing Board created under the *Ontario Farm Products Act* in 1947, and the LCBO, created under the *Liquor Control Act* in 1927. Both organizations mediate between clients and their suppliers, or in Lundvall's terms between users and producers of innovation. Each embodies historically rooted institutional attributes which are at variance with the evolution of the wine cluster. In the case of the OGGMB an adversarial relationship underlies the role of this organization as a mediator. Its mandate was to enable producers of a commodity to collectively market their product in an orderly manner and balance the power between many small individual sellers and relatively few larger buyers.

In the case of the LCBO, regulations governing the Ontario wine industry were shaped by immigrants from the British Isles and northern Europe who brought with them a Puritan ethic in which the consumption of alcohol topped the list of socially-disintegrating evils. Under the *Canadian Temperance Act of 1916*, however, wine had remained the only alcohol to be sold legally and small wineries proliferated. When prohibition was repealed, the government began restricting the growth of winemaking activities and the sale of wine. The LCBO developed laws to regulate alcohol consumption, operated only 16 stores throughout the province and imposed a long-term moratorium on the issuance of new winery licences. Over four decades, no new licences to produce and sell wine were granted in the province and a process of consolidation ensued in which the number of Ontario wineries fell from 61 in 1927 to six by 1974.[4] This would change in 1975 when the Liquor Licensing Board granted Don Ziraldo and Karl Kaiser the first winery licence since Prohibition. Inniskillin was created that year.

GROWTH AND TRANSFORMATION IN THE NIAGARA WINE CLUSTER

The Niagara Peninsula has been home to grape growing and winemaking for more than a century. Well into the 1970s this emerging cluster was based on local grape varieties, primarily Labruscas, Concords, and Niagaras, which made a sweet desert wine corresponding to earlier Canadian, and more broadly North American, tastes. In the midst of the Great Depression, one of Niagara's pioneering winemakers, T.G. Bright and Company, launched a long-term research and development program in both viticulture and viniculture. Convinced that European vinifera grapes could grow in the soil and under the climatic conditions of the Niagara Escarpment, Adhemar de Chaunac, Bright's French-born winemaker and chemist, succeeded in developing a large variety of hardy Canadian hybrids over the next 50 years. He was joined by other winemakers, grape growers, and agricultural research stations at Vineland (Ontario) and Somerland (British Columbia). Grape acreage planted to hybrids slowly expanded in Niagara and grape varieties such as de Chaunac, Foch, Baco Noir, Seyval Blanc and Vidal became the mainstays of a new table wine.

The creation of Inniskillin in 1975 signaled the relaxation of winery licensing regulations and led to the establishment of more commercial estate wineries in the late 1970s and early 1980s.[5] Many of the new estate wineries were founded by growers who had been experimenting with imported European vinifera grape varieties. Others were established by immigrants from Europe with the training in vini- and viticulture and were committed to quality winemaking. Together these actors spearheaded a movement toward premium wines made from vinifera grapes. While Labruscas continued to account for over 50 percent of the 17,887 acres planted to grapes in the Niagara district in 1981, hybrids had risen to 21.5 percent in that year and would reach almost 29 percent in 1986. Shortly thereafter hybrids joined labruscas in a steep and irrevocable decline (Table 1).

TABLE 1
Trends in the Evolution of the Niagara Wine Cluster

	1981	*1986*	*1991*	*1996*	*2002*
Grape Acreage					
Vinifera	459.8	446.6	1654.0	3402.8	7644.4
Hybrids	3851.1	5117.6	3688.9	2794.7	3818.2
Labrusca	10,071.2	9066.3	6656.4	5135.7	4359.5
Total	17,887.0	17,920.4	12,680.5	13,104.9	16,401.5
Grape Sales in Tonnes					
For wine	38,980	31,952	29,658	28,963	41,267
Total crop	69,925	77,240	56,630	47,738	53,186

Source: OGGMB *Annual Reports* (various years).

The transformation of the Niagara wine industry was precipitated by four factors. First, global consumption of dryer wines made from European vinifera varieties increased significantly: initially whites, such as Chardonnay, Riesling, and Sauvignon Blanc, more recently reds: Cabernet Sauvignon, Merlot, and Syrah. Second, in 1987 the General Agreement on Tariffs and Trade (GATT) decided to uphold a European Union complaint that the LCBO had engaged in discriminatory pricing practices. Third, Canada and the United States signed the Free Trade Agreement in 1988. Finally, new wine-producing countries, such as Australia, rapidly increased their exports to Canada.[6] The transformation was supported by government policies that funded the uprooting of older native and hybrid varieties and encouraged replanting with European viniferas. The impact of this policy has been dramatic. Vinifera plantings steadily increased from 2.5 percent of total grape acreage in the mid-1980s, to 13.1 percent in 1991, 26 percent in 1996, and 46.6 percent in 2001 (Table 1).

In 1989 the Vintner's Quality Alliance was put in place at the request of small estate wineries in order to establish quality standards for vinifera-based wines and thus build a more positive image for Canadian and Ontario wines amongst consumers. Products labelled with the VQA insignia would henceforth be made from 100 percent Ontario vinifera grapes grown in one of the three provincially designated viticultural areas: Niagara Peninsula, Lake Erie North Shore, or Pelee Island (OGGMB 2003).[7] VQA rules also stipulated labelling and content requirements. It is important to note, however, that adherence to VQA was strictly voluntary for the program's first ten years. Legal enforcement was not possible until the adoption of the *Vintners Quality Alliance Act* in 1999, which in 2001 vested authority in a new regulatory body, VQA Ontario.

While the content requirements and labelling of VQA wines were controlled by the VQA, non-VQA wines were governed by the Ontario *Wine Content Act* of 1973 and its subsequent amendments.[8] Since 1983, the act has allowed wineries producing non-VQA wine, primarily large wineries, to import up to 30 percent foreign material for blending in any bottle of wine. While large domestic wineries were satisfied with this rule, new wineries were pushing for change. One result of this push was the creation of the VQA and its subsequent transformation from a voluntary association into a regulatory body. When the Ontario *Wine Content Act* was renegotiated in 2001 many were disappointed that it still allowed older wineries, which were legally permitted to manufacture non-VQA wines, to import 30 percent foreign content for wine production. Once again rules were put in place that discriminated unfavourably against the small and emerging medium-sized wineries in the Niagara Peninsula.

Of the 70 to 80 wineries presently in Niagara, 69 are members of the Vintner's Quality Alliance, though not all wine produced by these wineries is VQA labelled. The peninsula continues to attract new wineries. Following the granting of three winery licences in the 1970s, eight wineries were established in the 1980s, 23 in the 1990s, and a further eight in the years 2000–03. Output of wine from the Niagara wine cluster has grown from 645,000 litres in 1991 to over 3.9 million litres in 2002 (Table A1).[9] The value of VQA wine sales has significantly increased over this period, from $5.6 million to more than $50 million (Table A2). Prior to the GATT decision, the LCBO had protected Ontario wines sales by placing a 60 percent markup on imported wines and only a 1 percent markup on domestic wines. Over the 1990s it eliminated this protection. Today, Ontario VQA wines are competitive in the domestic market and their sales account for nearly 30 percent of the domestic total.[10]

Despite this record of growth and innovation a number of disquieting trends have begun to appear. First, although the total acreage planted to grapes has not yet regained its level of 20 years ago and less than half the acreage is planted to viniferas, in 2003, Ontario had a surplus of vinifera grapes. Second, although the number of wineries has increased, most of these are small and even those established in the 1980s and early 1990s have not notably expanded. New large firms, moreover, are the offspring of consolidations and changes in marketing strategies. A glass ceiling seems to have been put in place with respect to the growth of smaller wineries. Third, although VQA wines are doing well in Ontario, the share of Ontario wines in total wine sales in the province has been declining over the past decade. In what follows we examine some of the factors that have contributed to these paradoxical trends.

THE ONTARIO GRAPE GROWER MARKETING BOARD

As defined by the Farm Products Marketing Commission, marketing boards are "producer elected, controlled, and financed. All producers of specified commodities are required to sell/market those commodities according to the marketing board's marketing plan" (OMAFRA 1998). Accordingly, the OGGMB is structured through an annual electoral process. Producers are divided into four districts: the Town of Niagara-on-the-Lake, the Town of St. Catharines and the District of Niagara South, the Town of Lincoln, and the Town of Grimsby and the Township of West Lincoln. Each grower has one vote, and all elected representatives must be grape growers. At their annual meeting, growers elect the Grape Growers Committee, with each committee member representing approximately 30 growers in their district. The Grape Growers Committee then elects seven directors, two from Niagara-on-the-Lake, one from St. Catharines, two from Lincoln, one from Grimsby and West Lincoln, and one director at large. The seven directors choose the officers for the board of directors, which approves the programs, policies and strategic direction of the Grape Growers of Ontario (OGGMB 2003). The board meets every month and conducts regular information and policy discussion meetings for growers throughout the year.

In its primary role as a marketing board, the OGGMB negotiates the floor price for all grapes sold to processors each year. Winemakers are represented by the Wine Council of Ontario and manufacturers of juice, jams and other grape products by the Ontario Food Processors Association. If prices cannot be settled through the negotiating process, the dispute is referred to an arbitration board.[11]

An estimated 80 percent of grape production goes into commercial winemaking and negotiations with the WCO are thus critical in setting the minimum prices for different grape varieties. Within these categories, prices vary according to grape quality, which the OGGMB measures solely in terms of sugar levels (brix). Between 1947 and 1974, consolidation in the grape-processing sector had posed a threat to the balance of market power between grape processors and growers. The OGGMB played an important role in balancing this power. In 2003, the Ontario Grape Growers Marketing Board was officially renamed the Grape Growers of Ontario, as the board felt that the term "marketing board" has a negative connotation and suggested that grape prices were being artificially manipulated (GGO 2003*b*). The structure and functions of the renamed organization, however, remained the same. The Grape Growers of Ontario is financed by growers, each of whom pays fees to the board for every metric tonne of grapes sold to processors. In addition to negotiating grape prices, the board lobbies for its membership, it facilitates information-sharing, and funds grape-related research (OGGMB 2003).

The 1980s and 1990s witnessed a transformation in Ontario's grape and wine sector. As opposed to six large wineries, uniform in size and market power, dominant in the sector from the 1950s into the 1980s, Ontario wineries are now highly differentiated by size, market power, products, and regulatory environment. The proliferation of small estate wineries in the 1990s has polarized the industry between a large number of small wineries and a few very large wineries (see Figure 2).

Ontario vintners now primarily differentiate wines according to non-VQA and VQA designations. Larger firms compete in all non-VQA and VQA market segments while small vintners predominantly focus on a narrow range of VQA market segments. This distinction has caused large and small wineries to enter into disagreements over grape-marketing regulations. Estate wineries are increasingly engaged in grape growing to meet the 1993 requirement that retail outlets must be located on at least five acres of vineyard.

During the 1970s and 1980s, a slow but steady decline in the number of Ontario grape growers began. Nonetheless, there remain about 530 grape growers in Ontario, with 397 engaged in wine grape production.

Toward the end of the 1980s and into the 1990s, a number of growers moved downstream in order to sell a higher value-added product: wine. Approximately 25 percent of all grape growers are currently vertically

TABLE 2

Grape Grower Demographics: Ontario Wine Grape Sellers 2002

Grapes Produced (tonnes)	*Number of Growers*	*Grape Production (%)*
>1000	4	14
500–1000	19	24
250–500	27	19
100–250	77	25
50–100	69	10
25–50	73	5
<25	128	3

Source: GGO *Annual Report* (2003).

FIGURE 2

Size of Wineries by Year of Establishment

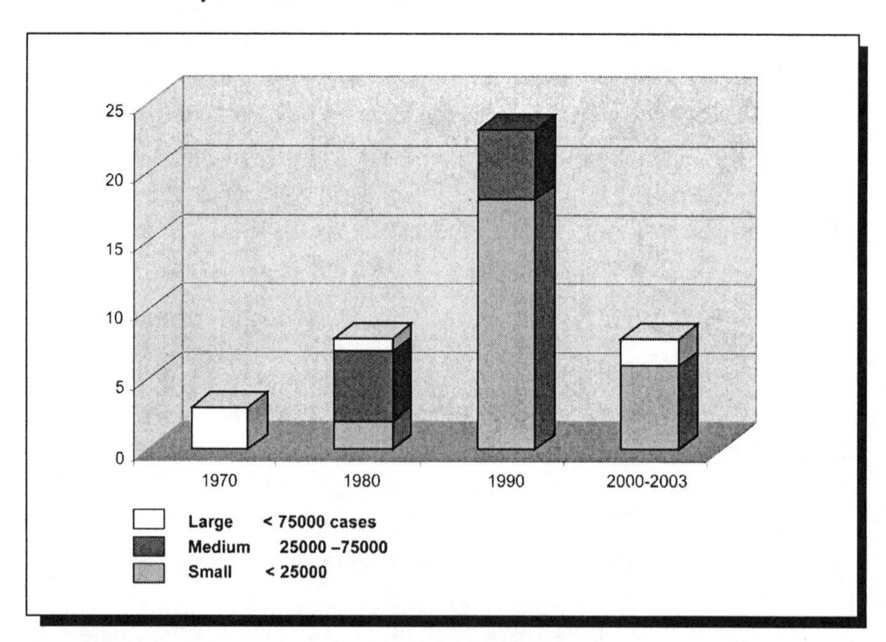

Notes: Includes 42 VQA wineries for which data were available. The large firms included: 1970, Inniskillin (Vincor), Chateau des Charmes (Independent), Hillebrand (Andres); 1980, Magnotta; 1990, none; 2000, Jackson Triggs (Vincor), Peller Estates (Andres).

Source: VQA Membership lists and interviews, authors' compilation.

integrated. Consequently, the distinction between wineries and growers is beginning to blur, with many estate wineries now characterized as grower-vintners. The rising number of grower-vintners confounds the initial purpose of the GGO in negotiating prices between small growers and large wineries. As both buyers and sellers of grapes, grower-vintners are doubly represented at annual grape price negotiations — by the GGO as grape growers and the Wine Council of Ontario as vintners. More importantly, being a grower-vintner also means that an estate winery pays marketing fees to the GGO in order to buy its own grapes. The paradox of grower-vintners positioned on both sides of the price negotiation table appears to render the organization redundant.

In essence, most grower-vintners do not require the GGO to act as a marketing intermediary. In fact, their costs would be lower if they did not have to pay marketing fees (called licensing fees). When the organization almost doubled the licensing fees for vinifera grapes in 2001, several grower-vintners refused to pay.[12]

TABLE 3
GGO Grape Licensing Fees
(per tonne)

Variety	*2000*	*2001*
Concord/Niagara	8.75	10.00
Vidal, Seyval Blanc	11.00	18.50
Experimental Red Hybrids	11.25	18.50
J. Riesling	15.00	29.00
Chardonnay	15.75	31.00
Sauvignon Blanc	16.25	29.25
Cabernet Sauvignon	19.25	32.75
Cabernet Franc	17.75	32.25
Merlot	19.50	33.00
Experimental Red Vinifera	16.25	29.50

Source: OGGMB *Annual Reports* (2000, 2001).

While the GGO reduced the fees the following year, and introduced structural changes to give grower-vintners a larger voice in the organization, the paradox of grower-vintners as buyer-sellers remains unresolved. Other issues have arisen from conflicting interests of VQA and non-VQA producers and from the recent vinifera surplus. These have renewed old tensions between growers and large wineries in particular. In 2001, the GGO reported that for the first time, vinifera production outpaced that of labruscas and French hybrids to the point of surplus. The surplus situation was repeated in 2002 as new grape plantings came into production, and is forecast to get worse each year (GGO 2001). The GGO has even gone so far as to urge growers to stop planting viniferas, as increased plantings will only depress prices (GGO 2003*b*).

One important power differential in the industry is the fact that the largest wineries buy 75 percent of the grapes grown on the peninsula. As such, many grape growers still see the Grape Growers of Ontario as an important buffer between them and the large players who dominate the grape market.

TABLE 4

Winery Demographics: Ontario Grape Buyers, 2002

Grape Tonnage Purchased	*Number of Wineries*	*Grape Purchases (%)*	*Grape Production as % of Purchases*
>1000 t	8	75.5	9.7
500–1000 t	7	11.1	54.1
100–500 t	18	9.6	62.4
25–100 t	25	3	170.5
<25 t	28	0.7	72.8

Source: GGO *Annual Report* (2003).

The *Wine Content Act* also creates division between large and small wineries. For non-VQA wines the act permits wineries to include a minimum of 75 percent Canadian grape content in wines in order to use the "Product of Canada" label and only 30 percent Canadian grape content for a "Cellared in Canada" label. This allows large wineries to import cheap foreign blending material from Chile, South Africa, and Australia, which hurts the market for Ontario grapes. Growers argue that this constitutes false advertising and prevents Canadian consumers from making informed choices about the wines they purchase. Imports of blending material, particularly by large wineries, have increased steadily since free trade despite the new emphasis on VQA in Ontario, while no wineries licensed after 1993 are legally permitted to import blending material. The practice of blending imported wine is particularly problematic in the light of vinifera grape surpluses in Ontario in 2000 and 2001.

TABLE 5
Trends in Grape Purchasing by Ontario Wineries

Grape Purchases in Ontario (tonnes)	*1989*	*1993*	*1997*	*2001*	*Growth 1989–2001 (%)*
Imported blending material	5,086	12,270	22,602	22,658	345
Ontario grapes	24,756	16,511	32,309	41,267	66
Total Ontario winery purchases	29,842	28,781	54,911	63,925	114

Source: OGGMB *Annual Reports* (various years).

Finally, while the growth of VQA output has been based on quality wine production on the Niagara Peninsula, grapes grown especially for VQA production still constitute a small portion of the grape supply. Grape growers are receiving mixed signals as to how to grow their grapes. In contrast to VQA production, the 30–75 percent domestic grape content for non-VQA products does not induce the same quality requirements for the mainly large vintners engaged in non-VQA production. Non-VQA production, which accounts for well over half of Ontario's grape demand, encourages a cheap price for grapes, which is countered by growers through higher yields.

VQA production, on the other hand, which accounts for a little over a quarter of the industry's grape demand, requires premium grapes grown at low yields at prices that both estate wineries and grape growers can manage. These factors are complex and largely negative for Ontario growers, and help to explain why the GGO has strengthened its commitment to VQA production significantly in recent years, while continuing to cling to its function as a market intermediary despite the grower-vintner dilemma.

Though the GGO is trying to adjust to the changes in the grape and wine sector, it has only recently recognized the burgeoning problems with the current pricing system. The GGO's *2003 Annual Report* indicates that the GGO is undertaking efforts to change its functions to better serve the interests of the Ontario grape and wine sector. In 2002, the Farm Products Marketing Commission hired a consultant to facilitate a discussion about grape pricing between the GGO and the Wine Council of Ontario. According to the GGO, the discussion revealed that the current pricing system "was not meeting the needs of both the growers or the wineries," and that as a result, "the GGO must make changes in its pricing methodology," while remaining committed to collective negotiations on behalf of growers (GGO 2003*a*, p. 1).

The price structure for Ontario grapes is based on brix (i.e., sugar content). The minimum grape prices negotiated by the GGO are based on a sliding sugar scale that calculates the weighted average brix content of the past five years. Since the 1980s, the weighted average brix level of grapes has increased substantially, supporting the GGO's belief that Canada's cool climate would not prevent growers from obtaining the sugar levels necessary for good wine grapes. Now that Ontario's grapes have attained acceptable brix levels, the bar of quality has once again been raised. However, on its own, brix is an insufficient measure of grape quality because acids and other complex compounds are critical to wine flavour. Two grape growers may produce grapes with identical brix levels, but due to differing viticulture techniques the varietal character of one bunch of grapes will be more complex than the other. In creating disincentives for innovation, the sliding sugar average incorporates every individual grape grower's innovations into the following year's base brix level without compensating the grower for his efforts, while ensuring that he or she must incorporate the same efforts in subsequent years in order to meet the base brix level. Moreover, acids and other flavours eventually decrease as brix increases. Quality is now a thorny issue for grape growers, but the fact remains that a price scale that equates grape quality with brix is problematic for innovative growers and wineries.

Second, the grape-price system is based on tonnage, which creates incentives for high volume output. Relationships between growers and vintners have changed as a result, as the top priority for vintners producing VQA wines has shifted beyond access to the right variety of grapes (i.e., viniferas) toward demand for higher quality viniferas. This has led to a focus on viticulture practices that produce low yields which allow for maximum flavour-development within the grape bunches. To compensate grape growers for losses incurred by low-yield techniques required to produce quality grapes, some vintners are now paying per acre prices. These special contractual arrangements are rare, however, given that integrated grower-vintners produce as much of their own premium grape content as possible to better control the quality of vineyard practices.

To reduce the risks that growers face when producing low-yield, quality viniferas, some vintners sign long-term contracts with preferred growers, which often create positive collaborative relationships in the vineyard between the contracted grower and the winery. Thus, while estate wineries are growing a portion of their own grapes, they maintain long-term contracts with trusted grape growers to ensure that their quality needs are met by purchasing a small quantity of grapes on the market for flexibility purposes.[13] Essentially, efforts to ensure premium quality within the small VQA segment of the wine grape market have been undertaken by vintners and growers negotiating amongst themselves, outside the GGO grape-pricing system. These arrangements indicate that the GGO is redundant for most innovative actors in the grape and wine sector. With respect to quality, the current price system creates more disincentives than synergies for innovation.

Besides the implications for relations between vintners and growers, long-term contracts are costly for vintners because grape prices in Ontario are higher than imported bulk wine from Chile, Australia, and Italy. Consequently, one very recent trend among larger vintners is consolidation and backward integration, through vineyard acquisitions or land purchases and new vineyard plantings of their own as they look to expand their VQA production. This strategy suggests that the GGO pricing system does not allow for grape-pricing flexibility along the quality spectrum, creating the incentives to acquire more vineyards in order to use different viticulture techniques for specific products (i.e., high yields for non-VQA content, medium yields for low-end VQA products like Vincor's Ancient Coast, and low yields for premium VQA wines). As these strategies are too costly for small estate vintners due to distribution problems at the LCBO, the only remaining option is to stay small and vertically integrated, and to compensate grape growers on the Niagara Peninsula for premium grape production.

The GGO's research and development (R&D) activities, largely limited to administration and joint-funding, are an attempt to counter the downsizing of the Horticultural Research Institute of Ontario at Vineland. Two joint-funded projects administered in the 1990s were the propagation of virus-free woodstock and the establishment of 12 automated weather stations in the Niagara Peninsula. In addition to these initiatives, the GGO provides a number of small research grants, mainly in the areas of pest and disease control. According to the organization, however, the board has become less focused on what needs

TABLE 6
GGO Promotion and Research Expenditures

GGO Expenditures	*1989*	*1993*	*1997*	*2001*
Promotion expenditures	149,963	187,003	142,651	188,520
Research expenditures	7,090	37,663	48,457	202,460
Total	520,557	615,035	585,526	976,375

Source: GGO *Annual Reports* (various years).

to be done in the areas of primary viticulture research, and as such has hired a consultant to assist in creating a long-term research vision and to implement a process to achieve these goals (GGO 2003*a*, p. 3).

The GGO has been slow to adapt to differentiating needs and demands of the Ontario wine and grape cluster. The continued dominance of large wineries in the grape-buying market suggests that its initial function may still be valid. However, the ability of wineries to compete, even in the domestic market, requires a continuous process of innovation in grape growing and not only in viniculture. Tensions over the pricing structure impede the development of trust-based interactions needed for innovativeness. Whether the GGO contributes to the dynamism of the wine sector will depend on how it deals with these tensions.

THE LIQUOR CONTROL BOARD OF ONTARIO

In the late 1980s, a series of domestic and international policy decisions strongly impacted Ontario's grape and wine industry and the policies and practices of the LCBO. In response to the General Agreement on Tariffs and Trade (GATT) decision in 1987 and the signing of a Canada-US Free Trade Agreement (FTA) in 1988, the LCBO eliminated the markup differential between domestic and imported wines, abolished import duties and began listing and distributing wine according to the country of origin. In accordance with the FTA provisions it also placed a moratorium on multiple retail licences. All wineries operating in 1988 were allowed to keep their winery retail stores, but all new winery licensees were only allowed one retail store at the winery site. This made the LCBO the sole provincewide distributor for all new wineries.

One other trend prompted a further change to liquor licensing. Between 1989 and 1993 Magnotta, Vinoteca, and Cilento began bottling foreign content wine within the city of Toronto. This enabled them to operate a retail outlet at the site and begin competing directly with the LCBO. In response, a 1993 amendment to the *Liquor License Act* required wineries to be located on at least five acres of vineyards in order to operate a retail store. In addition, the winery would have to carry out at least 25 percent by sales volume of the full primary fermentation process on site and the manufacturer of the remaining 75 percent would have to carry out at least one significant winemaking step. The licensing changes effectively made viticulture and winemaking activities a perquisite to wine retailing.

As the number of wineries within the wine cluster rose, a glass ceiling emerged with respect to the growth of smaller wineries. Three factors related to the operation of the LCBO contributed here: (i) the ability of small and medium-size wineries to access the Ontario market through the LCBO (getting listed), (ii) the LCBOs new marketing strategy and its effect on the ability of smaller wineries to stay listed, and (iii) the value-added of a provincial monopoly for the sale of wine and spirits which flowed from its substantial dividends to government, a large share of which came from imports.

Access to Market

The LCBO is the main vehicle for wineries to access the Ontario market. Only a few of Ontario's wineries have their own retail stores apart from the one store permitted at the vineyard: the four largest wineries, Vincor, Andres, Magnotta, and Chateau des Charmes and one medium-sized winery, Kittling Ridge. The top two have 270 private retail stores, the other three have 18 (Table 7).

These wineries also have a large share of the LCBO's listings, which has increased from 52.3 percent in 1998 to 66.8 percent in 2003 (Table 7). This resulted from an LCBO strategy to focus on volume sales, which led to a reduction in listings from 390 to 295, and from the practice of larger firms to increase their shelf space by buying up existing brands or introducing new ones. Vincor is a typical example. In addition to its older brands: Brights, Cartier, London, Inniskillin, Saw Mill Creek, and more recently Jackson-Triggs, the 2003 listings contained four new brands with a total of 12 new listings: Ancient Coast (five), Nero (three) and Sola (four). Between 1998 and 2003, listings

TABLE 7
Access to Markets, 2003

Winery	*Number of Private Retail Stores*	*Number and Share LCBO Table-Wine Listings*	
		1998 (N=390)	*2003 (N=295)*
Vincor	65 wine rack stores	100 (25.6%)	104 (35.3%)
Andres	105	73 (19.7%)	66 (22.4%)
Share of top two in total table wines		44.3%	57.7%
Percent share of top two in VQA wines		25 listings (23%)	38 listings (31%)
Percent share of top two in non-VQA wines		74% of 200 listings	76.3% of 173 listings
Magnotta	7	No listings (a)	No listings
Kittling Ridge	7	17 (4.4%)	14 (4.7%)
Chateau des Charmes	4	14 (3.6%)	15 (5.1%)
Total of these five wineries	288 stores	52.3% of total listings	66.8% of total listings

Note: LCBO did not list Magnotta while its court case was underway.

also rose for two middle-sized wineries: Cave Spring from five to seven and Henry of Pelham from seven to nine. But other mid-sized wineries, such as Pilliteri were delisted or the number of listings remained unchanged (Konzelmann at seven) or declined (Kittling Ridge from 17 to 14, Reif from six to four). Four smaller wineries — Semper (two), South Brook (one), Woods end (three) and York Hill (two) — were delisted and four other smaller wineries had their listings reduced in 2003: Peninsula Ridge (two), Strewn (two) Thirty Bench (one), and Willow Heights (one).

As a result of the tremendous growth in small estate wineries and low-volume VQA wine production throughout the mid- to late 1990s, the LCBO controls the main distribution network for Ontario wine. Yet there is a growing awareness that smaller wineries have not benefited from the existence of a state monopoly in this critical role. The LCBO has offered mixed responses to the situation, on the one hand arguing that their retail system encourages competition between smaller wineries, while on the other hand pointing to specific programs and "exceptions" made for Ontario wines despite Canada's trade obligations. A significant problem for smaller wineries lies in the LCBO's retail policies, in which product listings are based primarily on volume sales and secondarily on the store manager's discretion. Getting listed at the LCBO is itself a complex process and requires the development of a long-term strategy and marketing plan by the winery. This is a serious problem for small and medium enterprises that cannot spare personnel for strategic planning and marketing purposes and generally require assistance from business support services.[14] The process of getting listed is thus a lengthy one and the risks are all borne by the supplier if the product does not succeed.

The LCBO retail system has centralized merchandizing and distribution activities, the latter through six large warehouses that supply all 800 outlets. Nevertheless, there is considerable discretionary power left to local store managers in managing inventory. The LCBO system pushes products with the highest levels of volume sales system-wide, but below a certain sales performance level, the store manager has discretion to tailor store inventory to local consumer tastes. A new listing must achieve its annual provincial quota within 12 months from the announcement of its availability. As new products are reviewed quarterly, failure to achieve acceptable sales may lead to delisting before the 12 months are over. There are examples of early delisting, especially at the discretion of the local store manager. All risks incurred in delisting are borne by the supplier. When the LCBO delists a product the supplier must reimburse the LCBO for 25 percent of the purchase price and the LCBO can then sell off the remaining stock at a discount or the supplier takes back the remaining stock at their own expense.

Store size is another factor that shapes the quantity and selection of products available. The number of wine stock keeping units (SKUs) on the LCBO's General List amounts to about 2,000, with the largest stores offering about 1,600 wines from the General List and 1,500–2,000 from the Vintages list, a list of premium wines. Smaller stores, however, offer 250–300 wines from the General List and few, if any Vintage products. Agency stores operated by private retailers may sell as few as 50 wines, all from the General List.

TABLE 8
LCBO Store Breakdown

Number of Brands	*Number of LCBO Outlets*
>2,500	78
1,500–2,500	196
1,000–1,500	165
500–1,000	333
< 500	31
Total LCBO Outlets	803

Source: LCBO *Annual Report* (2001/2002, p. 54).

But discretionary practices of LCBO store managers is not the only problem. Despite the fact that Ontario table wines make up about two-thirds of domestic wine listings at the LCBO and 29 percent of total wine products on the General List, most Ontario VQA table wines are only available at a few LCBO stores. VQA wines are the stock-in-trade of the smaller estate wineries. This is particularly problematic in view of the vast array of wines from Australia, France, Italy, US, and Chile that are prevalent in most LCBO stores, and is related to the product-retail strategy employed in the majority of stores.

In general, the products that sell best get the most shelf space and are usually the most visible in the stores. The only exception is made for Ontario wines, which occupy desirable shelf space even when they are not always bestsellers. The LCBO has recently started to measure linear shelf footage, and points out that while Ontario wines account for 31 percent of sales, they have just under 33 percent of linear footage in most stores. The LCBO has committed to dedicating 35 percent of linear shelf space to Ontario wines by the end of fiscal year 2002–03. But most of the space is going to a small number of larger wineries and few of the smaller wineries benefit.

Recognizing this, the LCBO has created a number of programs to support smaller wineries such as the Hotline, Craft Winery, and Direct Delivery Programs. Through the Hotline program, wineries can ship directly to targeted

LCBO stores with more direct and preferential access to particular store managers. Wineries can therefore target specific markets for their products. The Destination Store is designed for unique low-volume premium products and unique varietals from both domestic and import suppliers. Intended for niche brands, selected wines are purchased in smaller quantities and are distributed to destination locations, approximately 20 LCBO stores and licensee depots through a forced distribution. Other LCBO stores are not permitted to stock these brands, enabling participating wineries to concentrate their promotional efforts exclusively on destination locations and growing estate wineries to develop and hone their in-store promotional activities. Wineries are expected to conduct product tastings to support sales. Some have criticized these programs, however, for only being available in a few southern Ontario stores.

The VQA Direct Delivery Program was introduced in 1999 in response to the decrease in independent winery stores sales after LCBO Sunday shopping was introduced. Before 1997, only winery retailers sold wine on Sundays. The decision to open on Sunday, made by the Ministry of Consumer and Business Affairs, increased the LCBO's level of competition with winery retail outlets, highlighting the LCBO's commitment to more aggressive retailing. As a result, winery retail stores suffered significant declines in sales, while the LCBO gained small sales increases. The VQA Direct Delivery program allows wineries to deliver their wines directly to bars and restaurants with only a 2 percent markup, constituting a revenue loss of $7 million to the LCBO.

The Craft Winery Program began in March 2002 as a three-year program to introduce products from "Growing Estate Wineries." Selected products can progress from a forced distribution to up to 60 predefined stores in the first year, to full General List at the end of the third year. The aim of this program is to enable wineries to develop brands for the General List, not for niche brands produced in limited production. This program gives small wineries premium shelf space usually exceeding their market share. By February 2003, only 11 small wineries had 17 wines on offer through this program. The craft winery program is thus not an antidote to the lack of listings, the difficulties of staying listed and the limited shelf space for small wineries.

Even for medium-sized wineries, the LCBO's product-management policy is a challenging one. Shelf placement depends on short-term sales performance; hence, if a product fails to meet the required sales quota, it is moved to a less prominent shelf where it is less noticeable to consumers. New wineries and new products are slow in getting onto LCBO shelves and have little consumer recognition by virtue of being new. They are nevertheless expected to meet the LCBO's provincial sales quotas within a matter of months. Combined with the Liquor Board's commitment to promoting wines from other countries, the task of staying on the shelf for both small and medium-sized wineries appears daunting.

In contrast to the situation of smaller wineries, LCBO product-management policies have not been a problem for larger wineries with diversified products. The LCBO is a significant distribution outlet for their more recognizable non-VQA products. As their non-VQA products came under pressure from cheaper imports, these firms made up for losses by expanding into VQA production (Table 7). In addition, sales from the retail stores of these five firms have increased by 47.1 percent over 1991 to 2002. This is despite the decrease in sales between 1998 and 1999 which occurred when the LCBO opened their stores on Sunday.

The LCBO Marketing Strategy: Staying Listed

Over the period 1988–2003, the LCBO was transformed from a state instrument designed to control the production and distribution of wine, into Ontario's leading large-scale marketer. In 1985, the LCBO launched its new image with a campaign entitled "Innovate, Merchandise & Generate Enthusiasm Programme (IMAGE)" designed as much for marketing as to encourage learning and adaptation within the organization. Changes within the LCBO included the creation in 1989 of a $1.7 million LCBO quality assurance laboratory, which many wineries came to depend on for testing. Beginning in 1991, the LCBO also began revamping its top 140 stores and launched a new campaign promoting Ontario wines. By 1994 all counter-service stores had been converted to self-service, and wineries were encouraged to promote products with in-store displays. Wine sales were facilitated by the growing use of debit and credit-card purchases and training programs for employees were launched.

In 1993, the LCBO renamed its *Today* magazine, *Food & Drink*, and began its "shop the world" marketing programs. Capital expenditures on marketing rose from $1.6 million in 1996, to $2.6 million in 1999. For smaller wineries, producing 100 percent VQA wines, the LCBO marketing strategy is highly risky because of the increasing costs of marketing programs and the uncertain and low returns. The LCBO, for example, charges suppliers for tastings, shelf-talkers, and extra promotional display materials. Promotional materials offered by the LCBO are sold to wineries, which gives larger wineries an advantage. These programs increased the LCBO's revenue and in 1995 the *Financial Post* declared that the LCBO was "Canada's most profitable company." In 1997 and 1998 the Retail Council of Canada named the LCBO "large-store innovative retailer of the year." The shift to high volume, rapid turnover marketing was completed with the building of mega-stores in Toronto (1999) and Ottawa (2000 and 2001). Between 1990–91 and 1999–2000 Canadian consumer tastes for alcoholic beverages shifted notably. Sales of spirits fell from 30.7 percent to 25.9 percent, sales of beer held steady at 51.7 percent, and sales of wines rose from 17.5 percent to 22.5 percent (Statistics Canada 2001, p. 4). At the same time consumers also increased their purchases of red wine relative to white. Ontario, which accounts for 40 percent of domestic consumption, experienced the strongest increase in red wine sales, up 17.3 percent over the period, much of which was imported. Despite already increasing sales of imported wine the LCBO launched a campaign to promote Australian wines in 2002. By volume imported, wine sales rose that year by 4.5 percent while domestic wine sales fell to just under 30 percent. That year Ontario sales of Australian wines increased by 22.3 percent, three times the growth rate of Chilean wines and 15 times that of US wines. LCBO sales figures also show, however, that sales of VQA wines have continued to grow, thus increasing their share of Ontario wine sales (Table A1). VQA wines have thus become strong competitors to imported wines. Their prices have fallen on average and they are able to increase their sales through the LCBO despite a nearly complete loss of markup protection. We are getting premium prices for premium wines despite the fact that VQA wines are not widely available in LCBO stores. Instead of promoting Australian wines from a state monopoly, one might have envisaged a different promotional campaign, something more along the lines of the 1991 promotion of Ontario wines, in which a two-for-one sale of matched Ontario and Australian wines was promoted as a vehicle for tasting and comparing to demonstrate just how far Ontario wines had come.

But another set of considerations might lead to a quite different strategy. The LCBO can count among its successes the substantial increases in the dividends paid to the provincial government through profit remittances, 72 percent of the total and the liquor sales tax. These rose from $793 million per year in 1993 to $1.2 billion in 2002. The federal government also received revenues from customs and excise taxes and the goods and services tax. These rose from $298 million in 1993, to $370 million in 2002. A recent study places

Ontario third in a field of 12 jurisdictions in wine levies per capita, behind Quebec at $8.40 and British Columbia at $7.17. This can be compared with Pennsylvania at $5.43, Washington State at $3.49, New Zealand at $2.29, and California at $1.13 (Girling & Associates 2003). The LCBO is thus a major contributor to provincial revenues. An important question is what proportion of these revenues is derived from imports. From a policy perspective, such considerations will need to be set against the relative employment, value-added, and environmental impacts of Ontario wine production as compared to wine imports.[15]

CONCLUSIONS

This chapter focused on learning in two organizations that have played an important role in promoting innovation and growth within the Niagara Peninsula wine cluster, the Grape Growers of Ontario, and the LCBO. Both organizations trace their establishment to the pre-World War II period and a set of conditions in the cluster and the industry that were significantly different from those that prevail today. Nonetheless, the first of these organizations, the GGO, has evolved little. Its underlying institutional base remains rooted in the notion that grape growers and vintners are distinct sets of actors whose interaction takes place mainly through a price-setting process that must be mediated not by the market but by a marketing board, an institution reminiscent of collective bargaining between actors with different and contradictory interests. A year ago, the OGGMB was renamed the Grape Growers of Ontario to counter claims that the prices fixed by the marketing board were out of line with the market. But nothing in this change of name implied a change in the adversarial nature of the rules governing the relationship between grape growers and vintners, despite the fact that these old habits and practices are now a disincentive to the longer-term collaboration between suppliers and clients that is often a positive factor in the growth of dynamic innovation systems (Mytelka and Farinelli 2003; Maskell and Malmberg 1999).

The second organization, the LCBO, dates to a still earlier period. But, unlike the GGO, the LCBO has undergone a profound transformation. From a learning perspective we might be tempted to see this as a positive example of adaptive learning. Current LCBO practices, however, are reducing the incentives for innovation and growth in small wineries. This is because the economic rewards to innovation for these wineries are constrained by the limited opportunities to amortize the costs of innovation through wider sales. These practices thus reinforce the negative signals inherent in the legal structure of wine retailing in the province. If there was a special role for a state monopoly in this area in the past, that role has been replaced by the LCBO's current retail policy and marketing strategy that, for all practical purposes, have turned the LCBO into a private retailer.

According to the Wine Council of Ontario, every $10 million in wine sales produces $16 million in economic activity in the province of Ontario (WCO 2001). With at least 10,000 hectares of good grape-growing land available for viniferas in the Niagara Peninsula, there is still room for growth. Although the Niagara wine cluster may be Canada's largest wine-producing region, Canada will remain a relatively small producer on the global scale relative to countries like Australia with its 130,600 hectares of bearing vineyards in 2001 (Zhou 2002, p. 14). But this should not discourage further expansion and growth of the cluster. For this to take place, however, the GGO and LCBO will need to rethink their roles.

NOTES

[1]Future papers will deal with a number of other issues related to the growth and innovativeness of the Niagara wine cluster such as land-use planning, research-industry linkages, and other interactive processes that form part of the dynamics of innovation in the cluster.

[2]The development literature provides numerous examples from Brazil and Korea, but similar practices have been used to stimulate the development of telecommunications equipment manufacture through the purchasing power of local telecom monopolies in a number of European countries.

[3]Curiously, some of our respondents exclude Guelph University but include the Vineland Station as an actor within the cluster. The interaction of Guelph University-based researchers with actors in the cluster — both researchers and vintners — has been a long and continuous one. From an innovation-system perspective, they are thus considered to be part of the system.

[4]These included Brights, Andres, Jordan, Cartier, and London.

[5]Inniskillin in 1975, Chateau des Charmes in 1978, Hillebrand (Newark) in 1979, Reif in 1982, Konzelmann in 1984, Stoney Ridge in 1985, Cave Spring in 1986 and still others in the late 1980s.

[6]The technological foundation for increased output in the Australian wine industry was laid in the 1960s and 1970s and accompanied by rising output in this period. "But the real take-off year can be dated quite precisely — it was 1986, when a big unexpected depreciation of the Australian dollar led to a surge of exports to the UK ...

and the proportion of output exported rose from about 3% in 1986–87 to about 30% in 1993–94." E-mail communication from Keith Smith, 3 April 2003.

[7]These areas are outlined in the *Vintners Quality Alliance Act*, 1999.

[8]The *Ontario Wine Content Act* falls under the Ontario Ministry of Consumer and Business Services. It has been traditionally renegotiated every five years.

[9]Compared to worldwide wine production, Canadian output is small: of the 267.6 hectolitres produced worldwide in 2001, Canada contributed 0.15 percent as compared with 19.9 percent in France, 19 percent in Italy, 5.9 percent in Argentina, 3.8 percent in Australia, 2.4 percent in South Africa and 2.1 percent in Chile (Leroux: 2003, p. 13).

[10]All data on wine production and sale are drawn from LCBO annual reports.

[11]*Farm Products Marketing Act*, Grapes for Processing, O.Reg.274/92,s.5.

[12]The licensing fees were subsequently reduced in 2002 to a more manageable level.

[13]Of the wineries interviewed, 11 indicated that this was the case.

[14]The United Nations Conference on Trade and Development (UNCTAD) maintains a small business support program, known as EMPRETEC in more than a dozen Latin American and African countries for this purpose. Within the European Union, to enable small firms to participate in R&D consortia, small business services and other intermediaries have similarly been used.

[15]The KPMG (2002) study has provided some indication of the relative employment and value-added impacts per litre of Ontario wine as compared to foreign wine sold on LCBO shelves. But more work is needed in this area.

REFERENCES

Aspler, T. 1993. *Vintage Canada: A Tasteful Companion to Canadian Wines.* Toronto: McGraw-Hill Ryerson.

Edquist, C. 1997. "Institutions and Organizations in Systems of Innovation: The State of the Art." Background paper for the Program Planning Workshop: "Strengthening Innovation Systems (SIS) An International Multi-Client Policy Development Program. Toronto, 26-27 June.

Girling & Associates Consulting. 2003. "Privatization Presentation." A study commissioned by the Wine Council of Ontario, 19 February.

Grape Growers of Ontario (GGO). 2001. *Crop Report.*

Grape Growers of Ontario. 2003*a*. *Annual Report.*

—— 2003*b*. *Newsletter*, January.

KPMG. 2002. "Study of the Ontario Economic Impact Content of Ontario Wines versus Imported Wines and the Ontario Economic Impact of Changes in the Relative Market Share of Ontario-Produced Wines. Toronto: KPMG, 29 May.

Leroux, G. 2003. "Ontario Grapes and Wines." Presentation to the Wine Council of Ontario, St. Catharines, 23 April.

Liquor Control Board of Ontario (LCBO). 2001/2002. *Annual Report*. Toronto: LCBO.

Lundvall B.-Å. 1988. "Innovation as an Interactive Process: From User-Producer Interaction to the National System of Innovation," in *Technical change and Economic Theory*, ed. G. Dosi, C. Freeman, R. Nelson, G. Silverberg and L. Soete. London: Pinter.

——, ed. 1992. *National Systems of Innovation: Towards a Theory of Innovation and Interactive Learning*. London Pinter.

Maskell, P. and A. Malmberg. 1999. "Localised Learning and Industrial Competitiveness," *Cambridge Journal of Economics* 23:167-85.

Mytelka, L.K. 2000. "Local Systems of Innovation in a Globalized World Economy," *Industry and Innovation* 7(1):33-54.

Mytelka, L.K. and F. Farinelli. 2003. "From Local Clusters to Innovation Systems," in *Systems of Innovation and Development: Evidence from Brazil*, ed. J.E. Cassiolato, H.M.M. Lastres and M.L. Maciel. London: Edward Elgar Publishers.

Nelson, R. 1993. *National Innovation Systems: A Comparative Analysis*. New York and Oxford: Oxford University Press.

Nelson, R. and S. Winter. 1982. *An Evolutionary Theory of Economic Change*. Cambridge, MA: Harvard University Press.

Ontario. Ministry of Agriculture, Food and Rural Affairs (OMAFRA). 1998. *Fact Sheet: Role of the Ontario Farm Products Marketing Commission in Ontario's Regulated Marketing System*. Toronto: Queen's Printer for Ontario.

Ontario Grape Growers Marketing Board (OGGMB). 2003. "About Us." Web site at <http://www.grapegrowersofontario.com/thegrowers/aboutus/index.html>.

—— various years. *Annual Reports*.

Statistics Canada. 2001. Cat. No. 11-001E, *The Daily*, 5 July.

Steward, Sean. 1997. "AGRIBUSINESS: Its Effects on the Niagara Grape and Wine Industry." A Senior Honours Thesis, Department of Geography, Faculty of Environmental Studies, University of Waterloo, April.

Storper, M. 1998. "Industrial Policy for Latecomers: Products, Conventions and Learning," in *Latecomers in the Global Economy*, ed. M. Storper, T. Thomadakis and L. Tsipouri. London: Routledge.

Wine Council of Ontario (WCO). 2001. *Wine Industry at a Glance*. St. Catharines, ON: WCO. At <http://www.vaxxine.com/wineroute/1facts.htm#economy>.

Zhao, X. 2002. "Who Bears the Burden and Who Receives the Gain?—The Case of GWRDC R&D Investments in the Australian Grape and Wine Industry." Melbourne: Department of Econometrics and Business Statistics, Monash University.

APPENDIX

TABLE A1

LCBO Total Wine Sales by Volume per Year, 1989–2002 (in '000s litres)

	1989	*1990*	*1991*	*1992*	*1993*	*1994*	*1995*
Domestic Wine (inc.VQA)	30,457*	26,002*	22,170	22,370	22,023	20,969	20,483
VQA wines	–	–	645	904	1,285	1,485	1,200
Domestic Wine Coolers	–	–	2,626	2,341	1,629	1,255	1,042
Imported Wine	44,922	45,476	45,814	44,017	43,966	44,951	46,375
Total Wine	**75,379**	**71,478**	**70,610**	**68,728**	**67,618**	**67,175**	**67,900**
Sales by ON winery stores	9,362	8,573	8,556	9,480	9,799	9,999	11,960
%Domestic/Total	–	–	31.6	32.5	32.6	31.2	30.2
% VQA/Domestic (inc.VQA)	–	–	2.9	4.0	5.8	7.1	5.9
% VQA/Total	–	–	0.9	1.3	1.9	2.2	1.8

	1996	*1997*	*1998*	*1999*	*2000*	*2001*	*2002*
Domestic Wine (inc.VQA)	22,208	24,957	25,845	25,795	26,523	26,958	26,934
VQA wines	1,660	2,494	2,534	2,718	3,158	3,659	3,933
Domestic Wine Coolers	903	895	718	544	489	499	549
Imported Wine	47,671	48,328	49,617	52,952	57,010	60,626	63,339
Total Wine	**70,782**	**74,180**	**76,180**	**79,291**	**84,022**	**88,083**	**90,822**
Sales by ON winery stores	13,164	14,411	14,838	13,878	14,074	14,961	16,164
%Domestic/Total	31.4	33.6	33.9	32.5	31.6	30.6	29.7
% VQA/Domestic (inc.VQA)	7.4	9.9	9.8	10.5	11.9	13.6	14.6
% VQA/Total	2.3	3.4	3.3	3.4	3.8	4.2	4.3

Note: *Canadian wines category includes both VQA and wine coolers in 1989 and 1990 figures.

APPENDIX

TABLE A2
LCBO Wine Sales by Net Value, 1989–2002 (in $'000s)

	1989	*1990*	*1991*	*1992*	*1993*	*1994*	*1995*
Domestic Wine(inc.VQA)	161,712*	150,080*	134,506	136,619	145,536	143,270	143,105
VQA wines	–	–	5,614	8,234	12,885	15,339	16,446
Domestic Wine Coolers	–	–	8,665	8,260	6,056	4,663	3,739
Imported Wine	367,652	385,041	396,053	372,925	379,358	390,427	410,573
Total Wine	**529,364**	**535,121**	**539,224**	**517,804**	**530,950**	**538,360**	**557,417**
Sales by ON winery stores	49,141	49,701	51,095	55,814	63,107	68,317	83,369
%Domestic/Total	30.5	28.0	24.9	26.4	27.4	26.6	25.7
% VQA/Domestic(inc.VQA)	–	–	4.2	6.0	8.9	10.7	11.5
% VQA/Total	–	–	1.0	1.6	2.4	2.8	3.0

	1996	*1997*	*1998*	*1999*	*2000*	*2001*	*2002*
Domestic Wine(inc.VQA)	158,888	178,563	190,809	201,723	211,595	214,443	217,234
VQA wines	20,075	27,671	31,146	34,827	41,259	47,770	50,205
Domestic Wine Coolers	3347	3,209	2,612	1,976	1,797	1,954	2,154
Imported Wine	435,845	469,803	503,290	559,624	635,112	680,993	731,605
Total Wine	**598,080**	**651,575**	**696,711**	**763,323**	**848,504**	**897,390**	**950,993**
Sales by ON winery stores	95,227	106,257	114,805	111,765	118,219	123,739	136,587
%Domestic/Total	26.6	27.4	27.4	26.4	24.9	23.9	22.8
% VQA/Domestic(inc.VQA)	12.6	15.5	16.3	17.3	19.5	22.3	23.1
% VQA/Total	3.4	4.2	4.5	4.6	4.9	5.3	5.3

Note: *Canadian wines category includes both VQA and wine coolers in 1989 and 1990 figures.

4

Biotechnology Companies and Clustering in Nova Scotia

Philip Rosson and Carolan McLarney

Introduction

This chapter focuses on an area of recent economic development, and one where there are high expectations for the future. The field of biotechnology has attracted increasing attention from scientists, business people, and government officials because breakthrough discoveries offer the prospect of improved social and economic conditions, greater wealth creation, and enhanced employment levels. Accordingly, many biotechnology start-ups and spinoffs have been established to apply science in areas including human health, food processing, agriculture, and the environment. Nova Scotia has witnessed considerable biotechnology activity: in 2003 there were approximately 28 companies operating in this field. Although small in absolute terms, the number is significant in a relatively small province. As our research is ongoing, this chapter provides a preliminary examination of biotechnology activity in Nova Scotia.

We begin with a brief literature review to frame the present study and then present some contextual material on biotechnology, including information on activity in Canada. The study is described in the third section, followed in the fourth section by preliminary findings. The latter contain baseline data on the companies that have been interviewed to date. Attention then turns to the issue of whether the companies in Nova Scotia presently constitute a biotechnology cluster. Clustering is examined from the standpoint of the companies interviewed, supplemented with information from organizations that support biotechnology development in various ways. In the final section we discuss and assess our findings.

THEORETICAL LITERATURE

Few firms operate in isolation, and dependence on others for business survival and growth is increasing. Whether the purpose is achieving economies through a supply chain, or facilitating innovation through joint research, interorganizational collaboration is assuming greater importance (Lorenzoni and Lipparini 1999). Consequently, researchers increasingly examine the linkages that exist between companies in dyadic relationships or more broadly in networks, such as the industry clusters that are of central interest in the research reported here.

Use of the cluster concept originated in the early twentieth century with the concept of agglomeration economies. Weber, Marshall and Schumpeter "suggested that firms locate together to reduce transaction costs, to increase flexibility and to achieve maximum information flow" (Hart n.d., p. 6). Interest in clusters mushroomed in the last 20 years as scholars, consultants, and economic developers attributed the economic growth of areas in the United States, Italy, Germany, and elsewhere to the emergence of specific industries in which the co-location of companies and support organizations had created a critical mass and dynamism (Saxenian 1994; Piore and Sabel 1984; Cooke and Morgan 1990). The recent interest in clusters also stems from Porter's influential work on the competitive advantage of countries, regions, and cities (Porter 1990, 1998*a*, *b*).

Many definitions of clusters exist. One simple view is that clusters are "groups of firms within one industry based in one geographical area" (Swann and Prevezer 1996, p. 139). This definition is extended by Porter when he states that clusters are "geographic concentrations of interconnected companies, specialized suppliers, service providers, firms in related industries, and associated institutions (for example, universities, standards agencies, and trade associations) in particular fields that compete but also cooperate" (Porter 1998*a*, p. 197).

It is widely regarded that firms benefit from being part of a cluster. Rosenfeld (2002) states that the interaction between companies — as well as support organizations — leads to "collective efficiencies." Malmberg and Maskell (2002) expand on this idea, stating that the main benefits of clusters are shared costs for infrastructure, the development of a skilled workforce, transaction efficiency, and knowledge spillovers that produce learning and innovation.[1] Proximity is critical to the realization of these benefits, particularly those that depend on face-to-face contacts and the transmission of tacit (as opposed to codified) knowledge. This is held to be an important consideration in advanced technology settings.

The present body of knowledge on industrial clusters is substantial, consisting of work conducted in different economic periods and geographic regions, and carried out by researchers from various traditions using contrasting methods. The resulting literature is rich but, at the same time, widely dispersed. The literature has benefited from review and synthesis efforts. Such contributions always provoke new questions and this review is no exception. We examine three of these contributions to provide a sense of the issues that have been raised. Martin and Sunley (2003) have identified important questions about cluster *definitions*. They argue that many key cluster elements, such as geographical and industrial boundaries, linkage strength, and proximity measures, are imprecisely defined. A second issue concerns cluster *types*. Rather than a single cluster model, field studies have found a variety of configurations, prompting the development of several cluster typologies. The best-known typology is that of Markusen (1996), but others have also been proposed (Camagni 1991; Gordon and McCann 2000; Hart n.d.). This raises the possibility that other configurations also exist. Questions have been asked about the *performance* of clusters. Staber (2001) finds that much of the research on clusters is anecdotal and descriptive. Martin and Sunley (2003) believe that studies do not demonstrate that firms perform better within clusters. More extensive research is required, particularly studies that compare similar firms inside and outside clusters. These examples give a sense of the challenges facing cluster researchers. Some of these scholars feel that the "rush to employ 'cluster ideas' has run ahead of many fundamental conceptual, theoretical and empirical questions" (Martin and Sunley 2003, p. 5).

We now turn our attention to innovation. Firms increasingly view networks as an important source of, and vehicle for, innovation (Lundgren 1993). Studies show, for example, that the number of research and development (R&D) linkages between companies is growing sharply (Hotz-Hart 2000), and that firms that cooperate intensely are more innovative than those that do not (Smith 1995). While collaboration is important for firms of all sizes (Lorenzoni and Baden-Fuller 1995), it is particularly so for the resource-constrained, smaller companies that account for a disproportionate share of new, breakthrough technology. Clusters potentially have much to offer such firms, including a pool of skilled workers, the transfer of tacit knowledge, input from key customers and/or suppliers, and access to venture capital funds. In fact, research shows that new and small firms frequently capitalize on spillover knowledge from other firms and research institutions, especially during early stages in the industry life cycle (Audretsch 2003). Smaller firms contribute to the innovative ability of clusters in several ways. Their size permits them to "move quickly and flexibly combine, break-up and recombine resources in different ways" (Cook and Pandit 2002, p. 230). In addition, the high failure rate of small businesses often enhances learning and knowledge diffusion in clusters, when released employees move from one local company to another (Longhi 1999).

BIOTECHNOLOGY

Biotechnology is not a single industry but rather a scientific knowledge base that is changing sectors such as pharmaceuticals, medical devices, agriculture, and the environment. Based on the application of recombinant DNA and its related technologies, biotechnology dates back to the 1983 patent applications of Cohen and Boyer, scientists at Stanford University and the University of California at San Francisco respectively. These scientists discovered a way to move genes between organisms and, in so doing, turned the science of molecular biology into commercially useful knowledge. Because biotechnology involves radically new technologies that often require long discovery and development lead times and face an uncertain regulatory approvals process, the financial demands are significant. Large expenditures are required to get to the point where a discovery can be patented, and even greater resources are needed to bring the resulting product to market. These conditions are problematic for the small and medium-sized companies which tend to innovate in this area. It is thus not unusual for innovating companies to enter into alliances with, or be acquired by, larger, established industry players.

Biotechnology discoveries have transformed a variety of industries: biotechnology is estimated to produce global revenues of $500 billion annually and growth rates of 20 percent per year (Nova Scotia 2000). Consequently, many countries view biotechnology as a way to drive economic growth, and governments are investing in science and technology in the hope of achieving significant returns. Such investment takes many forms, including funding research institutions and programs, establishing strategies to support local company growth, encouraging inward investment, and seeding the development of industry clusters.

Biotechnology activity is concentrated in space, as demonstrated in recent US and European studies (Cooke 2003; Cortright and Mayer 2002; Yarkin 2000). Although large cities such as Boston and Cambridge are dominant forces in biotechnology, clusters are also found in smaller metropolitan centres and in locations outside major industrial areas. For example, although the Paris region dominates French biotechnology (accounting for one-half of the companies), five other regions have established vibrant biotechnology clusters on the periphery of the country (Mytelka 2001).[2] The same holds in Canada, where the cities of Toronto, Montreal, and Vancouver are home to most biotechnology firms. At the same time, smaller agglomerations are found in cities such as Saskatoon and Edmonton (Niosi and Bas 1999).

THE CANADIAN CONTEXT

Biotechnology in Canada

The biotechnology sector in Canada is substantial and growing. Canada is home to the second-largest number of biotechnology companies in the world (after the US) and ranks third in revenues generated (after the United States and United Kingdom). Metrics such as market capitalization and R&D spending — as well as number of firms and revenues — show strong growth over the last five years (Ernst & Young 2002). There were 358 biotechnology innovator firms in Canada in 2001.[3] The human health sector (199) accounted for more than half (53 percent), with agriculture (67), food processing (46), and environment (31) also prominent. The companies were concentrated in three Canadian provinces: Quebec (130 or 35 percent), Ontario (101 or 27 percent), and British Columbia (69 or 18 percent). Most were small in size: 267 (or 71 percent) had 50 or fewer employees, 62 had 51–150 employees, and 46 companies had 151 employees or more. Revenues grew by 343% between 1997 and 2001 (to $3.5 billion), reflecting greater levels of commercialization of R&D. However, the industry is "young" and many companies have still to generate any revenues from their biotechnology research (Statistics Canada 2003).

The Case of Nova Scotia

Because biotechnology is not an industry with its own set of SIC codes, it is difficult to be precise about the number of biotechnology companies that exist in any given jurisdiction. Statistics Canada (2003) identified 12 biotechnology innovator firms in Nova Scotia in 2001 (or 3 percent of the Canadian total). Through a variety of methods (see below), we identified a total of 28 companies involved in biotechnology at the end of 2002. Whether the number of companies is 12 or 28, Nova Scotia is clearly a minor biotechnology player in national terms. However, biotechnology contributes substantially to the province's economy and it is believed that potential exists for further development.

It was estimated that in 2000, biotechnology research in the Halifax Regional Municipality accounted for (i) spending of over $86 million annually, (ii) employment of 2,340 professional researchers and technicians, (iii) annual salaries of $124 million, and (iv) provincial government tax revenues of $25 million (LSDA 2002).[4] Much of the research activity is centred at Dalhousie University, whose medical school has identified neuroscience, infectious diseases, cardiovascular, and cancer research as core areas of competence. A second significant research organization is the Institute of Marine Biosciences (IMB) at the National Research Council, which has research expertise in the fields of microbial genomics and marine biosciences. The IMB is located adjacent to the Dalhousie University campus. These organizations have produced spinoff companies and collaborate closely with local companies. Two other public research institutes are located within the Halifax metropolitan area. The Bedford Institute of Oceanography is part of the federal Department of Fisheries and Oceans and its 300 researchers conduct research on ocean science. InNOVAcorp is a provincial Crown corporation that conducts research in several fields and operates incubator facilities that house approximately ten biotechnology companies.

A number of other organizations play an important role. Governments at the federal and provincial levels have identified biotechnology as a priority sector (in Nova Scotia, and more particularly Halifax) and one that should therefore be supported through a variety of programs. Biotechnology is one of three sectors emphasized by the Greater Halifax Partnership, which is a private sector-led economic development organization. Two industry associations are also present. BioNova was founded in 1993 and provides leadership, advocacy, information, and education services to about 40 members, including biotechnology firms, support organizations, and service firms. It is located in a life sciences incubator (one of those run by InNOVAcorp). The Life Sciences Development Association (LSDA) was established in 2001. Its principal goal is to bring state-of-the-art research facilities to Halifax. The LSDA is located within one city block of Dalhousie's medical school. Aside from various forms of business funding provided by the federal and provincial governments, competitive grants are provided by Genome Atlantic, a national not-for-profit organization established in 1999 with federal government money. Two venture capital companies are located in Halifax. MedInnova Partners was set up by the Canadian Medical Discoveries Fund to help develop early-stage technologies. It exists to bridge the funding gap between research grants and the marketplace and has made several local investments. ACF Equity is the second venture capitalist. To date it has made no investments in biotechnology companies.

THE STUDY

Fieldwork

This chapter is based on information collected through interviews with 33 biotechnology companies and support organizations between December 2002 and September 2003. A database of biotechnology companies and support organizations in Nova Scotia was constructed using lists from government and industry associations, and entries in business directories (e.g., *Canadian Biotechnology, Industry & Suppliers Guide 2002*). The database was expanded as a result of suggestions made as the interviews proceeded. Our interest is in biotechnology activity within 100 kilometres of Halifax and so contact was restricted to those lying within this boundary.[5] Initial contact was made by mail with a phone call follow-up shortly thereafter. Of 28 companies identified, 14 have been interviewed, six remain to be interviewed, five did not reply or refused to participate, two companies had ceased operations or moved back into a university laboratory, and one had moved to Ontario.[6] Interviews have been conducted in 19 support organizations, including research institutes, government agencies, industry associations, and venture capitalists. All of the support organizations agreed to participate in the study and these interviews are essentially completed.

The interviews were semi-structured. An interview guide was used to collect the desired information but, when deemed necessary, additional questions were asked. Five types of interview guides were employed (company, research institute, government agency, industry association, and venture capitalist) which included some common and other more particular questions.[7]

Interviews, ranging in duration from one to two hours, were usually conducted on-site, although in a few cases the meeting was held at Dalhousie University. All interviews were recorded and later transcribed.

Nova Scotia Biotechnology Companies: Baseline Data

Table 1 presents a snapshot of biotechnology companies in Nova Scotia in 2003; this will be expanded as the study is completed. The names of the companies in question have been withheld for confidentiality reasons. Several comments can be made about the data presented. Of the 14 companies listed, three were established in the 1980s, eight in the 1990s, and three since 2000. While a range of scientific approaches are evident, most of the companies (11) are focused on human health applications. Two of the companies, I and K,[8] have workforces of 150 or more and may be regarded as medium-sized. The majority, however, are very small; eight have ten or fewer employees. The small scale of the companies is also reflected by their revenues and R&D expenditures. Eight report having revenue streams, five currently have zero revenues, and another company (M) would not provide data. The greatest annual sales were achieved by the three longest established companies ($13 million, $10 million, and $5 million). Four companies are spending in excess of $1 million annually on R&D activities. In terms of development stage, eight of the companies have commercial products that generate cash flows.[9] Three others (B, L, and N) are at the clinical trials stage, while three (A, G, and H) are at an earlier stage of development. Reflecting their development stage, five companies have annual R&D expenditures that presently exceed revenues. Finally, 12 of the 14 companies are privately owned.

The companies exhibit similarities as well as differences. The similarities are that these are small biotechnology enterprises that have predominantly targeted human health applications and are funded privately. The differences chiefly relate to the stage of business development and the precise focus of their products and research efforts. A number of companies have stood the test of time and now have products, revenue streams, and an operational base. Others have yet to achieve sustainability. Some have modest operations which help to finance continuing research efforts, while the remainder are exclusively focused on research in the laboratory, consuming rather than producing cash at the present time. The human health interests of many of the firms are diverse in terms of scientific method and application.

TABLE 1

Profile of Nova Scotia Biotechnology Companies Interviewed

Company	*Start-up*	*Ownership*	*Employees*	*Revenues ($m)*	*R&D ($m)*	*Stage of Development*	*Focus*
A	2000	Private	1	0	–	1	Drug and vaccine delivery system
B	1995	Public	37	<0.50	1.20	3.3	Rapid diagnostic testing
C	1999	Private	3	<0.50	<0.10	4	Medical device
D	2001	Private	8	<0.50	<0.10	4	Biomedicinals to health-care professionals and consumers
E	1998	Private	2	<0.50	<0.10	4	Oceanographic instruments
F	1993	Public	41	0	10.00	4/2	Active pharma ingredients
G	1999	Private	4	0	0.37	1	Gene research for neurological conditions
H	2000	Private	10	0	0.40	2	Vaccines for human and veterinary applications
I	1987	Private	150	10.00	4.00	4	Marine-based nutraceuticals
J	1983	Private	22	5.00	0.05	4	Frozen human plasmas
K	1981	Private	150–300	13.00	0.50	4	Products derived from seaweeds
L	1993	Private	7	0	1.30	3.3	Applications from chitosan
M	1997	Private	16	–	0.30	4	Tele-health applications
N	1997	Private	5	<0.50	0.18	3.2	Gene mapping for animal reproductivity longevity

Notes:

1. Financial data are for most recent year.
2. Stage of development: 1 = development, 2 = pre-clinical, 3.1 = initial clinical trials, 3.2 = animal trials, 3.3 = human trials, 4 = commercialization, – = not provided.

A Biotechnology Cluster?

In this section, we provide a partial analysis of the data collected to date. As noted above, we examine the Nova Scotia biotechnology companies from a cluster perspective. We do this by posing five questions: (i) How important are local connections to biotechnology companies? (ii) Which factors have contributed most to local growth in biotechnology? (iii) Which factors have impeded local growth? (iv) Are there service providers in Halifax that cater to biotechnology companies? (v) Does a biotechnology cluster exist in Halifax?

How important are local connections to biotechnology companies? Four connection measures were selected for analysis and are shown in Table 2. These were local customers, local suppliers, local knowledge relationships, and local financing. As noted above, "local" is defined as being located within 100 kilometres of Halifax. Only three companies (C, D, and K) relied to any degree (30 percent, 10 percent, and 5 percent respectively) on proximate

TABLE 2

Importance of Local Connections for Nova Scotia Biotechnology Companies

Company	*Local Customers (%)*	*Local Suppliers (%)*	*Local Knowledge Relationships (%)*	*Local Financing (%)*
A	n/a	0	0	100
B	0	50	20	0
C	30	80	60	100
D	10	20	0	20
E	0	10	0	20
F	0	15	20	0
G	n/a	0	50	100
H	n/a	0	50	80
I	minimal	30	major	100
J	minimal	20	major	100
K	5	50	50	100
L	n/a	40	90	66
M	0	0	90	100
N	0	40	90	0

Note: Read as follows: 30 percent of the customers of C are located within 100 kilometres of the company; n/a = not applicable.

markets. Six companies (E, F, I, and J) reported that their customers were not local, but located in other parts of Canada, the US, or in international markets. Four companies (A, G, H, and L) do not currently have commercial products available for sale and thus do not have customers at present. Finally, company F is presently a captive supplier to its US parent but expects to broaden its customer base with the development of a new product in the near future.

When we examine connectedness to *suppliers*, a range of arrangements is evident. At one extreme, four companies (A, G, H, and M) source nothing locally, whereas company C relies on local suppliers for 80 percent of its inputs. Others (B and K) split their buying equally between local and non-local suppliers, while two more (L and N) depend somewhat less on the local market for supplies. Local suppliers account for 30 percent or less of the inputs for the remainder (D, E, F, I, and J).

Knowledge relationships include a variety of modes through which companies develop their know-how and expertise: collaborative research projects, personnel exchanges and secondments, research consortia participation, public research invention licensing/patenting, and new technology development or adoption. Eleven of the 14 companies reported that to some degree they are involved locally in knowledge relationships. In three cases (L, M, and N), these were substantial, accounting for 90 percent of the firm's knowledge relationships. Two more companies (I and J) are also involved locally in a major way. In four more cases (C, G, H, and K), half or more of the companies' knowledge relationships were local. Finally, two companies (B and F) had quite modest relationships (20 percent) and three (A, D, and E) had no local knowledge relationships at all.

The final measure of local connectedness is the extent to which companies accessed local *financing*. This is a major issue for companies and is raised again below. Seven of the companies (A, C, G, I, J, K, and M) indicated that 100 percent of their financing was from local sources, including local venture capital firms, angel investors, and government. Two other companies (H and L) receive two-thirds or more of their funding locally, whereas for firms D and E the figure is 20 percent. Company F is a subsidiary of a US corporation, which fully funds its operations. Firms B and N are also financed totally from afar.

These data show that local connections vary by company and by the specific measure used. Company C reports the greatest degree of connectedness across all four measures, whereas companies E and F appear to be least embedded in the local area. Three other companies (I, J, and K) show relatively high levels of local knowledge relationship and financing links as well as some supply connections. However, their local customer connections are of minor concern. Regarding the four measures of connection, companies appear to depend most on the local area for financing, followed by knowledge relationships. Supply linkages operate at a considerably lower level and customer connections are by some margin the least important benefit provided by the local area to Nova Scotia biotechnology companies.

Which factors have contributed most to local growth in biotechnology? Companies were asked to comment on eight factors as potential contributors to the growth of the local biotechnology industry. Table 3 shows that in total, six of the factors were endorsed. "Specialized research institutes and universities" was most frequently regarded as important (11 companies). This is not surprising given that Dalhousie University researchers were prominent in the founding and development of several of the firms. The medical school was particularly prominent, but some of the companies had their origins in biology and oceanography departments. Acadia University and the Nova

TABLE 3
Factors Contributing to Nova Scotia Biotechnology Growth

Company	*Co-location*	*Supply of Workers*	*Infrastructures*	*Research Institutions*	*Training Institutions*	*Government Policies*
A				x		x
B	x	x	x	x		
C				x		x
D	x	x	x	x		x
E	x		x	x	x	x
F		x	x		x	x
G	x	x				
H	x	x	x	x		
I		x	x	x	x	x
J		x		x		x
K			x	x		x
L		x		x		x
M		x		x		
N					x	
Total	5	9	7	11	4	9

Scotia Agricultural College were instrumental in the development of the science that is the basis for other companies. Nine firms selected "supply of workers with particular skills" as a growth factor. In general, companies reported satisfaction with the supply of technicians from university and college programs. This was regarded as of continuing importance as more companies move toward manufacturing operations. "Government policies and programs" were also felt to be beneficial by a majority of biotechnology firms. "Physical, transportation or communications infrastructures" (seven companies), "co-location with other firms in the same industry" (five companies) and "specialized training or educational institutions" (four companies) received the balance of support. In no case did a company endorse "availability of financing" or "presence of key suppliers and/or customers" as a factor in biotechnology growth in Nova Scotia.

Which factors have impeded local growth in biotechnology? The same eight factors were presented to companies as potentially having inhibited biotechnology growth in Nova Scotia. Table 4 summarizes the responses of the 14

TABLE 4

Factors Inhibiting Nova Scotia Biotechnology Growth

Company	*Co-location*	*Supply of Workers*	*Infrastructures*	*Research Institutions*	*Availability of Financing*	*Government Policies*
A		x			x	
B					x	
C					x	
D		x	x		x	
E					x	
F					x	
G	x	x			x	
H		x	x		x	
I					x	
J	x	x			x	
K		x				
L						
M			x		x	x
N	x	x		x	x	x
Total	3	7	3	1	12	2

companies. All but two companies chose "availability of financing" as a growth inhibitor. Most firms complained that there is insufficient venture capital available and many have turned to angel investors for their funding. Financing is seen as an area in need of pressing attention if biotechnology is to develop a more significant presence in Nova Scotia. "Supply of workers" was identified by seven companies and is the only other factor to be noted by more than a few firms. This seems paradoxical in that companies cited the same factor as a reason for growth. The explanation here is that companies have been able to find skilled technicians for their operations, but have experienced problems in recruiting senior managers with ten years of relevant experience. Ideally, the experience would be in biotechnology but, as a minimum, should be with a science- or technology-based firm. Local companies have filled the resulting management gap through retired executives who have re-settled in Nova Scotia. "Co-location" and "infrastructures" (three each), "government policies" (two) and "research institutions" (one) were also endorsed as inhibitors to growth by a small number of companies. These were also cited as growth factors and so some commentary is needed. The conflicting messages about the benefits of co-location are attributable to the fact that while firms generally view their proximity to related firms positively, the diversity of products and operations means that there are few if any tangible complementarities. Other factors simply reflect the fact that some companies are more demanding than others and/or have had varying experiences. Finally, one company (L) was more general in its comments, viewing Nova Scotia as too small and lacking the resources and sophistication to be attractive as a biotechnology venue at the present time.

Are there service providers in Halifax that cater to biotechnology companies? One indicator of the development of an industry is the presence of specialized service providers that cater to its members. Local accountants and lawyers are used by almost all the 14 companies and are able to deal with most business matters in a satisfactory manner. When an issue cannot be resolved locally, it is usually referred to another branch of the accounting or law firm or, failing that, to an associated company. One issue, however, is problematic — the matter of patent law. While local patent agents/investigators exist in Halifax and companies report using their services, the same is not true for biotechnology patent expertise. At present, companies depend on patent lawyers in Ottawa, New York, and San Diego.

Does a biotechnology cluster exist in Halifax? The previous four questions provide indirect evidence about the presence of a biotechnology cluster in Halifax. The final question dealt with the matter directly. Contrasting opinions emerged when companies were asked, "Do you consider your company to be part of a network of related firms in your region/locality, i.e., a cluster?" Five companies answered "yes" while nine said "no." A few companies spoke about feeling connected to institutions such as Dalhousie University and the National Research Council rather than to other firms.

Most respondents regarded the industry in Nova Scotia at present as being at an early stage in its development and, as a result, to consist of a collection of biotechnology companies rather than comprising a "cluster." Biotechnology support organizations were also sceptical that a cluster exists at this time. Repeated mention was made of the fact that although a group of companies exists in biotechnology, their small size and recent establishment meant that most were preoccupied with research and commercialization. Consequently, most firms are inward focused and have little time for interaction with others in the proximate area. Their diverse interests also rule out much chance for synergies. In addition, some companies and support organizations expressed the view that an anchor firm was needed to provide the necessary focus and leadership for cluster development. At present, however, there is no semblance of a core set of technologies or interests in the area that might attract the attention of an anchor firm.

DISCUSSION

In this final section, the findings reported above are discussed in relation to widely held views about clusters. These findings are then linked to recent research on clusters and their development.

Preliminary Findings

The cluster literature stresses the helpful presence (and sometimes demanding requirements) of upstream and downstream actors. However, analysis of the preliminary interview data suggests that biotechnology companies in Nova Scotia depend relatively little on the local area as a source of inputs, and even less as an outlet for their products. Knowledge relationships with local research institutes and universities are evident but, given other researchers' findings on biotechnology growth (e.g., Niosi and Bas 1999), closer ties might have been expected. Nonetheless, local institutes do provide a secure "home" for some of the company scientists and have provided streams of capable technicians and scientists to Nova Scotia firms. The local area does play a more significant role in the financing of Halifax companies. This is perhaps to be expected given the rule of thumb that venture capitalists prefer to be within an hour of their clients. Halifax is about two hours by air from Toronto, Ottawa, and Montreal, and about the same from Boston and New York. Thus, local sources of finance, from private investors and venture capitalists, are important.

Companies commented on eight factors that are widely regarded as affecting location decisions (and clustering).[10] The factors cited most often as having facilitated biotechnology growth were research institutes, government programs, and supply of workers (technical). The factors that have inhibited growth most were stated to be availability of financing and supply of workers (managerial). Some companies viewed infrastructures and co-location as having helped growth, whereas others expressed the opposite view. These responses suggest that Nova Scotia has both strengths and weaknesses as a location for biotechnology firms in general, and for a cluster in particular. Related to this point, the company responses show that at the present time the biotechnology industry is too recent and small to support specialized service providers, particularly patent lawyers. Finally, a majority of firms and support organizations do not view Halifax as currently having a "true" biotechnology cluster. In summary, although there seems to be a sizeable number of companies and organizations, these do not presently constitute a cluster — at least as described in the mainstream literature, where an industry cluster is seen to be richly endowed and characterized by intense exchanges that take place between firms and support organizations.

A Broader Context

The above results suggest that Nova Scotia does not presently have a biotechnology cluster. It is worth examining this question a little further by viewing the present findings in a broader context. We first raise the question of local versus non-local connections before moving on to consider the cluster-development process.

The relative lack of local connections has led us to question whether Nova Scotia biotechnology firms constitute a cluster. Recent studies present a rather more complex picture of interactions in clusters than was true in the past. For example, Bathelt, Malmberg and Maskell (2002) argue that the connections of cluster members can be simultaneously close to home and extending around the world (thus producing "local buzz" and tapping into "global pipelines"). Other researchers also conclude that cluster members need extra-local links if they are to learn and remain competitive in a global business environment (Simmie 2002). This is particularly the case in biotechnology, where there is a widely distributed knowledge base (Cooke 2001). Canadian (Gertler and Levitte 2003) and Swedish researchers (Coenen *et al.* 2003; Dahlander and McKelvey 2003) demonstrate the importance of global as well as local interactions to support biotechnology innovation.[11] Taken together, these research findings suggest that the cluster may have been overemphasized as a knowledge source in earlier research. This somewhat different viewpoint might explain the Nova Scotia findings, which reveal that non-local customers and suppliers, and to a lesser extent knowledge sources, are important to biotechnology firms.

The Nova Scotia findings should also be interpreted with respect to time. According to traditional views, it seems that Nova Scotia does not presently have a biotechnology cluster. But perhaps it is unreasonable to expect that it might. While the major research institutions were established many years ago and some biotechnology activity dates back to the 1980s, the majority of firms got their start in the 1990s. As several studies attest, time is an important factor in the development of a cluster. "Building a technology cluster is the work of decades. To the outside world in the mid-1990s, Ottawa looked like an overnight success story. But the groundwork had actually been laid 20, 30, 40 and even 50 years earlier … 'You have to build a nucleus around something deep-rooted'" (Mallett 2002, p. 12). These comments about a Canadian information technology cluster are echoed elsewhere. Waluszewski (2003) finds that the biotechnology cluster in Uppsala developed over many years and not, as some believe, as a direct consequence of the 1996 restructuring of a major pharmaceutical company. These studies reinforce the importance of evolutionary processes that occur over decades and as a result of interaction between many actors. Consequently, the more relevant question might be "Is there potential for a biotechnology cluster in Nova Scotia?" rather than "Does a biotechnology cluster presently exist?"

The answer to this question must wait until the final interviews have taken place and more extensive data analysis is completed. In the meantime, because Nova Scotia lies outside the industrial heartland of Canada, it is worth noting work that has examined innovation and cluster development in such regions. Rosenfeld identifies the barriers that exist in such regions as well as actions that can be taken.[12] The barriers include "a weak infrastructure; lack of access to technology, innovation and capital; regional insularity and isolation; low educational levels and low skilled work force; absence of talent; and an overly mature or hierarchical industry structure" (Rosenfeld 2002, p. 9). The company responses reported above indicate that some of these barriers are present in Nova Scotia. Others are not, however, suggesting that there is an opportunity for innovation and cluster development. Developments in other jurisdictions are informative. For example, Feldman (2002) discusses the development of dynamic clusters in biotechnology and information technology in the Capital region of the US, an area that was not known for entrepreneurship or high technology. This example provides evidence that patient and adaptive cluster-development initiatives can produce results. This is an important area for further research because business development in less industrialized regions faces considerable challenges. Yet these regions account for the majority of locations. As Dicken and Malmberg comment, "Not all spatial agglomerations of similar or related industrial activity are 'Hollywoods' or 'Silicon Valleys'" (2001, p. 13).

NOTES

[1]A number of writers extend the analysis further, listing factors that lead to cluster growth and *decline* (Cook and Pandit 2002, p. 223).

[2]Technopoles play an important role, bringing together academic institutions, government laboratories, companies, and financial service firms in each location.

[3]An "innovative biotechnology firm" is one that uses biotechnology for developing new products or processes and is engaged in biotechnology related R&D activities.

[4]Halifax is the capital of Nova Scotia and home to 40 percent of its people. With a population of 360,000 in 2001, Halifax is Canada's thirteenth largest metropolitan area.

[5]Following the lead of others, the SSHRC-funded studies all use 100 kilometres to define the limits of a local area. Almost all biotechnology activity in Nova Scotia takes place within 100 kilometres of Halifax.

[6]The move to Mississauga was required under a venture capital funding arrangement.

[7]These same interview guides are being used in all of the case studies that make up the research project.

[8]Company K is shown as having 150–300 employees. The larger number reflects seasonal employment required for harvesting seaweed.

[9]Company F manufactures a commercial product for its parent and is in the process of developing a new formulation.

[10]Another location question was put to the companies. When asked "Why is your company located in this region/locality?" the overwhelming response was that it was due to personal circumstance or preference. The founder, owner, and/or principals lived in the community in question and were unable or unwilling to move. In two cases, employment was also mentioned. Specifically, the business owners were also employed as university scientists and presently juggled both responsibilities. A few other firms mentioned secondary location factors such as the presence of research and training institutions and government support programs.

[11]The need for frequent and intense exchanges with other cluster members is also challenged by Fontes' (2003) research, which describes how Portuguese biotechnology companies are able to access required knowledge *without* belonging to a cluster.

[12]Rosenfeld (2002) identifies three types of less-developed region: (i) older industrialized regions that have lost their cost advantage to other nations, (ii) semi-industrialized regions with small, craft-based firms with low levels of technology, and (iii) peripheral or less populated regions where productivity and out-migration concerns mean that resource-based industries must be supplemented by those with growth opportunities.

REFERENCES

Audretsch, D. 2003. "Innovation and Spatial Externalities," *International Regional Science Review* 26(2):167-74.

Bathelt, H., A. Malmberg and P. Maskell. 2002. "Clusters and Knowledge: Local Buzz, Global Pipelines and the Process of Knowledge Creation," DRUID Working Paper No. 02–12. Copenhagen Business School. At <www.druid.dk/wp/pdf_files/02-12.pdf>.

Camagni, R. 1991. "Local 'Milieu,' Uncertainty and Innovation Networks: Towards a New Dynamic Theory of Economic Space," in *Innovation Networks: Spatial Perspectives*, ed. R. Camagni. London: Belhaven, pp. 121-42.

Canadian Biotechnology, Industry & Suppliers Guide 2002. Georgetown, ON: Contact Canada.

Coenen, L., J. Moodysson, B. Asheim and O. Jonsson. 2003. "The Role of Proximities for Knowledge Dynamics in a Cross-Border Region: Biotechnology in Øresund." Paper presented at the DRUID Summer Conference 2003, Copenhagen. At <www.druid.dk/conferences>.

Cook, G. and N. Pandit. 2002. "Innovation, Small Firms and Clustering: Insights from the British Broadcasting Industry," in *New Technology-based Firms in the New Millenium*, ed. R. Oakey, W. During and S. Kauser. Oxford: Pergamon.

Cooke, P. 2001. "New Economy Innovation Systems: Biotechnology in Europe and the USA," *Industry and Innovation* 8(3):267-89.

—— 2003. "Regional Science Policy? The Rationale from Biosciences," in *Clusters: Old and New*, ed. D. Wolfe. Kingston and Montreal: School of Policy Studies, Queen's University and McGill-Queen's University Press.

Cooke, P. and K. Morgan. 1990. *Learning Through Networking: Regional Innovation and the Lessons of Baden-Wurttemberg*, Regional Industrial Research Report No. 5. Cardiff: University of Wales.

Cortright, J. and H. Mayer. 2002. *Signs of Life: The Growth of Biotechnology Centers in the US*. Washington, DC: The Brookings Institution, Center on Urban and Metropolitan Policy.

Dahlander, L. and M. McKelvey. 2003. "Revisiting Frequency and Spatial Distribution: Innovation Collaboration in Biotech Firms." Paper presented at the DRUID Summer Conference 2003, Copenhagen. At <www.druid.dk/conferences>.

Dicken, P. and A. Malmberg. 2001. "Firms in Territories: A Relational Perspective," *Economic Geography* 77(4):345-63.

Ernst & Young. 2002. *Beyond Borders: The Canadian Biotechnology Report 2002*. Toronto: Ernst & Young.

Feldman, M. 2002. "The Entrepreneurial Event Revisited: An Examination of New Firm Formation in the Regional Context," *Industrial and Corporate Change* 10:861-91

Fontes, M. 2003. "Distant Networking: The Knowledge Acquisition Strategies of 'Out-Cluster' Biotechnology Firms." Paper presented at the DRUID Summer Conference 2003, Copenhagen. At <www.druid.dk/conferences>.

Gertler, M. and Y. Levitte. 2003. "Local Nodes in Global Networks: The Geography of Knowledge Flows in Biotechnology Innovation." Paper presented at the DRUID Summer Conference 2003, Copenhagen. At <www.druid.dk/conferences>.

Gordon, I. and P. McCann. 2000. "Industrial Clusters: Complexes, Agglomeration and/or Social Networks," *Urban Studies* 37(3):513-32.

Hart, D. n.d. "Innovation Clusters: Key Concepts." Unpublished paper, University of Reading. At <www.rdg.ac.uk/LM/LM/fulltxt/0600.pdf>.

Hotz-Hart, B. 2000. "Innovation Networks, Regions, and Globalization," in *The Oxford Handbook of Economic Geography*, ed., G. Clark, M. Feldman and M. Gertler. Oxford: Oxford University Press.

Life Sciences Development Association (LSDA). 2002. *Strategy for the Commercialization of Life Sciences Research.* Halifax, NS: Life Sciences Development Association.

Longhi, C. 1999. "Networks, Collective Learning and Technology Development in Innovative High Technology Regions: The Case of Sophia-Antopolis," *Regional Studies* 33:333-42.

Lorenzoni, G. and A. Lipparini. 1999. "The Leveraging of Interfirm Relationships as a Distinctive Organizational Capability: A Longitudinal Study," *Strategic Management Journal* 20:317-38.

Lorenzoni, G. and C. Baden-Fuller. 1995. "Creating a Strategic Center to Manage a Web of Partners," *California Management Review* 37(3):146-63.

Lundgren, A. 1993. "Technological Innovation and the Emergence and Evolution of Industrial Networks: The Case of Digital Image Technology in Sweden," *Advances in Marketing* 5:145-70.

Mallett, J. 2002. *Silicon Valley North: The Formation of the Ottawa Innovation Cluster*. Ottawa: Information Technology Association of Canada.

Malmberg, A. and P. Maskell. 2002. "The Elusive Concept of Localization Economies: Towards a Knowledge-based Theory of Spatial Clustering," *Environment and Planning* 34:429-49.

Markusen, A. 1996. "Sticky Places in Slippery Space. A Typology of Industrial Districts," *Economic Geography* 72:293-313.

Martin, R. and P. Sunley. 2003. "Deconstructing Clusters: Chaotic Concept or Policy Panacea?" *Journal of Economic Geography* 3:5-35.

Mytelka, L. 2001. "Clustering, Long Distance Partnerships and the SME: A Study of the French Biotechnology Sector." At <www.utoronto.ca/isrn/documents/Mytelka_Clustering%20Long%20Distance.pdf>.

Niosi, J. and T. Bas. 1999. "The Competencies of Regions: Canada's Clusters in Biotechnology," *Small Business Economics* 17(1):31-42.

Nova Scotia. 2000. *Opportunities for Prosperity: A New Economic Growth Strategy for Nova Scotians*. Halifax, NS: Government of Nova Scotia.

Piore, M. and C. Sabel. 1984. *The Second Industrial Divide*. New York: Free Press.

Porter, M. 1990. *The Competitive Advantage of Nations*. London: Macmillan.

—— 1998*a*. *On Competition*. Cambridge, MA: Harvard Business School Press.

—— 1998*b*. "Clusters and the New Economics of Competitiveness," *Harvard Business Review* (December):77-90.

Rosenfeld, S. 2001. "Backing into Clusters: Retrofitting Public Policies." Paper presented at Symposium on Integration Pressures: Lessons from Around the World, Cambridge, Harvard University. At <www.rtsinc.org>.

—— 2002. *Creating Smart Systems: A Guide to Cluster Strategies in Less Favoured Regions*. April. At <www.rtsinc.org>.

Saxenian, A. 1994. *Regional Advantage: Culture and Competition in Silicon Valley and Route 128*. Cambridge, MA: Harvard University Press.

Simmie, J. 2002. "Knowledge Spillovers and the Reasons for the Concentration of Innovative SMEs," *Urban Studies* 39(5/6):885-902.

Smith, K. 1995. "Interactions in Knowledge Systems: Foundations, Policy Implications and Empirical Methods," *STI Review* 16:69-102.

Swann, G. and M. Prevezer. 1996. "A Comparison of the Dynamics of Industrial Clustering in Computing and Biotechnology," in *The Dynamics of Industrial Clustering: International Comparisons in Computing and Biotechnology*, ed., G. Swann, M. Prevezer and D. Stout. Oxford: Oxford University Press.

Staber, U. 2001. "The Structure of Networks in Industrial Districts," *International Journal of Urban and Regional Research* 25(3):537-52.

Statistics Canada. 2003. "Distribution of Biotechnology Innovator Firms in Canada, 2001," and "Coming of Age: Biotech Revenues Are on the Rise," *Innovation Analysis Bulletin* 5(1):10-14.

Waluszewski, A. 2003. "What's Behind a Prospering Biotech Valley? A Competing or Cooperating Cluster or Seven Decades of Combinatory Resources?" Department of Business Studies, Uppsala University, July 7. Unpublished paper.

Yarkin, C. 2000. "Assessing the Role of the University of California in the State's Biotechnology Economy," in *The Economic and Social Dynamics of Biotechnology*, ed. J. de la Mothe and J. Niosi. Boston: Kluwer Academic Publishers, pp. 115-31.

5

THE BIOTECHNOLOGY CLUSTER IN VANCOUVER

J. Adam Holbrook, M. Salazar, N. Crowden, S. Reibling, K. Warfield and N. Weiner

INTRODUCTION

A firm's ability to innovate, which is one of the most powerful sources of competitive advantage in modern economies, is determined by its capacity to acquire, adapt and advance knowledge. Knowledge is a unique commodity because it can be created, but not destroyed, and when it is transferred the source retains all of the knowledge it transfers to the recipient. Knowledge can flow from one institution to another either through transfers of people or through financial transactions that permit the acquisition of knowledge.

Technology-based clusters are emerging in regions that have achieved critical mass in the knowledge economy. Strong research universities, industrial laboratories, and entrepreneurial companies, with human capital and infrastructure, anchor these clusters. Collectively these clusters form regional and national innovation systems. This chapter examines one such cluster: the biotechnology industry in Vancouver.

The literature on national innovation systems (NIS) is quite recent. Christopher Freeman introduced the concept in his 1987 study of Japan and Lundvall developed it further in his 1992 study of Denmark. Since then the systems of innovation approach has broadened its perspective to include regional systems of innovation (RIS). This raises the question of whether a national system is greater than the sum of its regional parts. The focus on spatial issues has two major aspects; on the one hand, it recognizes that innovation is a *social process* and is shaped by persons and institutions that share a common language, rules, norms, and culture (i.e., common modes of communication and practice). On the other hand, innovation is also a *geographic process*, taking into account that technological capabilities are grounded in regional communities that share a common knowledge base.

A national system of innovation is more complex in a federation such as Canada than in a centrally administered nation because provincial/state and national institutions and actors share power. Canada is one of the few true economic, social, and political federations in the world. In the Organisation for Economic Co-operation and Development (OECD), only Australia, the United States and Germany have socio-economic and political features similar to Canada. Thus, Canada's NIS is different from that in most other nations. A key element of the Canadian federation is the allocation of most economic powers to the national government and most social responsibilities, particularly health and education, to the provinces.

Holbrook and Wolfe (2000) have argued that, at least in the case of Canada, in order to understand the NIS one must first understand the RIS. This task is complicated by the wide variations in geography, population, and economy that distinguish the regional systems. National level data of the Canadian system of innovation are heavily biased by the economic activities occurring in the two major industrialized provinces, Ontario and Quebec. In most developed nations, the central government formulates science and technology policies, yet most innovation activities take place locally. Thus, nationwide innovation policies may not affect each region equally, and could be counterproductive in some instances.

The recent OECD territorial review on Canada recommends a new regional policy approach that requires the

> valorization of potential competitive advantages with regard to industrial production and services, and the removal of bottlenecks (weak cluster integration, valorization of natural resources, etc.) preventing further development. Given their often local and regional nature, this strategy should result in attaching a more important role to the territorial policies (OECD 2002).

There are many empirical studies on regional innovation systems. Some have developed different typologies of RIS and clusters to highlight complex variations that exist within and between countries. Longhi (1998) argues that the development of a regional innovation system requires three conditions:

(i) a coherent set of territorial relationships among all economic actors; (ii) a specific culture; and (iii) a shared representation system, implying a strong consensus and integration among them. Cooke (1998) identifies two key aspects of RIS: the Science and Technology (S&T) research governance infrastructure and business innovation superstructure (see Table 1). Cooke uses the governance infrastructure dimension to classify modes of technology transfer. The business innovation dimension gives the posture of the firms in the regional economy, both toward each other and the outside world, as well as in relations with producers and consumers in the marketplace.

TABLE 1
Some Examples of Regional Innovation Systems

Governance Structure/ Business Innovation Dimension	*Grassroots*	*Network*	*Dirigiste*
Localist	Tuscany (northern Italian industrial districts)	Tampere (Denmark)	Tohoku (Japan)
Interactive	Catalonia *Saskatchewan* *Manitoba*	Baden-Wurtemberg *British Columbia* *Alberta*	Quebec
Globalized	Ontario California	North Rhine-Westphalia	Singapore Midi-Pyrenées

Source: Cooke (1998). Regions in italics are added by the author.

The governance infrastructure dimension, initially developed for technology-transfer purposes, establishes three main types of RIS: grassroots, network, and *dirigiste*. It is clear that governance structure for Cooke is not political governance as it is usually understood. The initiation of the RIS is the key feature, which then affects funding, the type of research (applied, basic, near to the market, etc.), technology specialization, and the forms and degrees of coordination. Grassroots RIS are locally organized, network RIS are multi-level organized and *dirigiste* RIS are the product of central government policies.

Going a step forward, one needs to distinguish between an RIS and an industrial cluster of the type defined by Porter. How much innovation and what type of innovation should exist in a cluster for it to be considered a viable element of an RIS? Several cluster definitions exist:

- Clusters are geographic concentrations of interconnected companies and institutions in a particular field (Porter 1998).
- A cluster is a geographically bounded concentration of interdependent businesses (Rosenfeld 1997, cited by Asheim and Isaksen 2002).
- Clusters are regarded as places where close inter-firm communication, and social-cultural structures and institutional environments may stimulate socially and territorially embedded collective learning and continuous innovation (Asheim and Isaksen 2002).

THE BRITISH COLUMBIA BIOTECH CLUSTER STUDY

There are a number of potential biotech clusters in Canada, seven of which are being studied by the ISRN project. The long-term objective is to compare these clusters within the NIS and identify regional differences that can provide useful information about the RIS. For the purposes of this study, we started with a definition of a cluster based on Porter's model, but modified as described in the overall ISRN program description.[1]

Building on work by Peter Phillips, Wolfe and Gertler distinguish between two types of industrial clusters in Canada: type I, "regionally embedded and anchored," and type II, "entrepôt." In a type I cluster, "the local knowledge/science base represents a major generator of new, unique knowledge assets." In a type II cluster "much of the knowledge required for innovation and production is simply acquired through straightforward market transactions" (2003, pp. 28-29). Table 2 shows the clusters and a first approximation of their typology. In the cases of the three smaller clusters, it is difficult to determine typology. They are shown here as "undifferentiated." It may well be that at some point they will grow to a point that they become either a type I or a type II cluster. Less likely, but still a possibility, is that type II clusters eventually grow to be type I. The ISRN research should help to determine this.

Not all of these studies have been completed at this time and this chapter highlights the biotech cluster in the province of British Columbia. The definition of "biotechnology" is that used by the OECD (2001):

> The application of S&T to living organisms as well as parts, products and models thereof, to alter living or non-living materials for the production of knowledge, goods and services.

It includes the following categories:

1. DNA (the coding): genomics, pharmaco-genetics, gene probes, DNA sequencing/synthesis/amplification, genetic engineering.
2. Proteins and molecules (the functional blocks): protein/peptide sequencing/synthesis, lipid/protein engineering, proteomics, hormones, and growth factors, cell receptors/signalling/pheromones.
3. Cell and tissue culture and engineering: cell/tissue culture, tissue engineering, hybridization, cellular fusion, vaccine/immune stimulants, embryo manipulation.
4. Process biotechnologies: Bio-reactors, fermentation, bio-processing, bioleaching, bio-pulping, bio-bleaching, bio-desulphurization, bioremediation, and bio-filtration.
5. Sub-cellular organisms: gene therapy, viral vectors.

It should be noted that this definition excludes medical devices based on biotechnological processes. The initial work was based on firms and institutions based in Vancouver; in later years the hypothesis that the cluster extends to

TABLE 2
Typology of Biotech Clusters in Canada

Cluster	*Type*	*Size ("in stars")**	*Geographic Boundaries*	*Economic Boundaries*
Montreal	I	70	++	++
Toronto	I	47	–	++
Vancouver	I	80	++	++
Saskatoon	II	22	Wide area	+
Ottawa	undifferentiated	6	+	+
London	undifferentiated	5	–	– –
Halifax	undifferentiated	n/a	++	+

Note: *See Queenton and Niosi (2003).

firms on Vancouver Island will be tested, but it is not clear yet whether this assumption is valid. For practical reasons, as well as for the geographic concentration usually associated with industrial clusters, this study does not extend to other parts of British Columbia.

The institutions and enterprises making up the cluster for our study are taken from a "snapshot" in early 2002 based on information from the National Research Council (NRC) and BC Biotech.[2] There are approximately 40 privately owned firms, ten venture capitalists, nine government organizations, two non-profit organizations, and three research institutes. Thirty-two private firms belong to the pharmaceutical cluster and eight to the medical device cluster. This number is constantly changing: BC Biotech (2002) stated that there are 91 biotech firms in BC in 2002 with 88 of them in the Vancouver/Lower Vancouver Island area.

THE BIOTECH "VIBE" IN VANCOUVER

Biotech research institutes in Vancouver are, understandably, located near the three main postsecondary institutions: the University of British Columbia (UBC), Simon Fraser University (SFU), and the British Columbia Institute of Technology. Government agencies are located on or near the UBC campus.

The private component of the biotech sector in Vancouver consists of young and small firms. The oldest (and "model") is Quadra Logic Technology (QLT) founded in 1981, which is the largest privately owned biotech firm in Vancouver. Since 1995, 19 biotech firms have been spun off from UBC, and seven from SFU; 19 of these (73 percent) are still in existence. Figure 1 shows the growth of the BC biotech cluster start-up firms by year of start-up and stage of development.

The firms are located in three very narrowly defined neighbourhoods — the UBC campus, the Vancouver General Hospital, and the Burnaby/New Westminster industrial area. All but one of the venture capitalists are located in the financial district in downtown Vancouver. An interesting observation is that the firms often "trade up" from one lab facility to the next, by taking over larger premises and releasing their old space to newer, smaller firms. At the same time many of the specialized facilities built to incubate biotech companies (often financed with public money and located in public institutions) are now empty as they are too expensive for start-up companies.

FIGURE 1
Growth of Vancouver Biotech Cluster

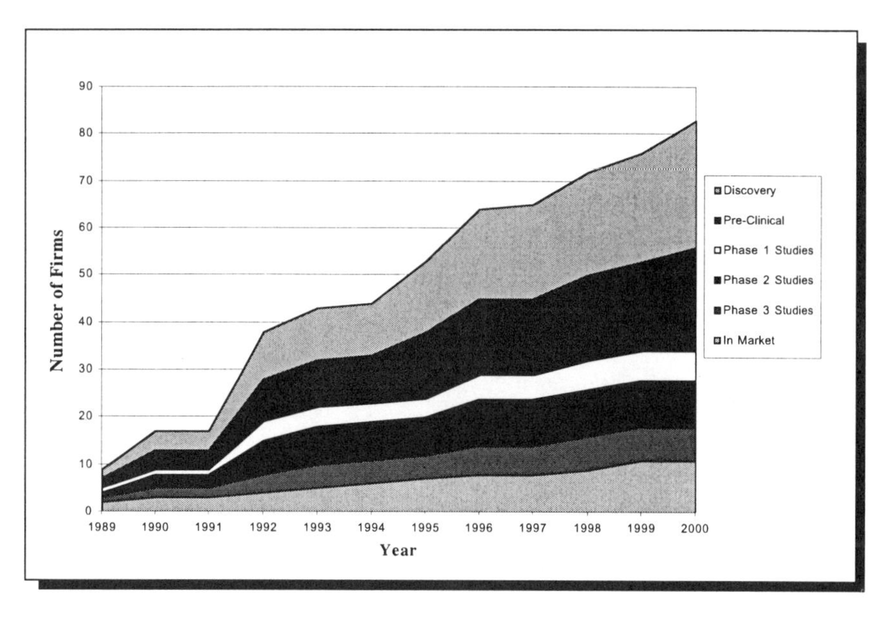

BARRIERS TO INNOVATION

In discussing barriers to innovation, respondents reported difficulty finding and retaining qualified employees. There is substantial competition with the US and eastern Canada due to economic disadvantages in Vancouver, primarily high personal income taxes and housing costs. These are somewhat offset by the cultural and climatic conditions of the area. As some respondents noted,

> "Need some kind of incentive to attract employees, something we can work with other than scenery – like tax free stock options…"
>
> "No incentive to finance biotech in Canada."
>
> "Best thing government could do is leave us alone."
>
> "Let us do business without interfering."
>
> "Lack of experienced management, upper level."

Observers of the biotech cluster (such as government officials) argued that the lack of industrial experience in Vancouver is due to the relative youth of the cluster. The lack of highly qualified and experienced management is a problem, not only in Vancouver but also across Canada. Apart from a lack of senior management personnel, there is a lack of expertise in the regulatory/drug development area. Respondents have reported the existence of "prowling" head hunters, but inter-firm movement by technical personnel is relatively low. When asked what could be done to overcome these barriers, responses included,

> "Tax breaks! Like Quebec!"
>
> "Tax credits (or tax holidays) for upper level management to be attracted to work here."
>
> "Finance – more grants (no pay back)."
>
> "Need to increase science coming out of universities. Not enough commercially minded scientists."

It was argued that it is difficult to attract upper-level personnel to Vancouver because of its relatively remote location on the North American continent. There are not enough large firms, and potential executives feel that there is not enough scope for mobility among biotech enterprises.[3] Additionally individuals perceive that British Columbia and Canada are high-tax locations and that Canadian salaries are not competitive with US salaries; however, once recruited to Vancouver they are less likely to leave.[4]

SOURCES OF INNOVATION

When asked about sources of innovation, respondents noted that there is now a constant progression of trained PhD students from UBC labs to contribute to the talent pool in BC. According to one, "it's a trickle down effect better than in '92." Innovation is helped by the close proximity of firms to university labs: "opportunities come from UBC, SFU, and the University of Victoria where tech comes from university and genome centres." The biotechnology cluster appears to have grown from knowledge spillovers coming out of the universities (Goldfarb and Henrekson 2003).

There is a strong, collegial biotech community in Vancouver, even though there are no business relations between these companies. Many companies claim that they are not "part of a network of related firms in the region." Most claim to work in a specific niche, developing a particular technology/product primarily for global pharmaceutical companies. It seems that there are two biotech specialities in Vancouver, based on the OECD definition and the categories it provides to classify the industry: DNA-genomics, and proteins and molecules (especially proteomics, and peptides sequencing and synthesis). The situation appears to be similar for service companies and contract research organizations. Often they were established originally to support the development of the local cluster, but their actual customers are all around the world, especially US companies.

There is a strong network of biotech innovators: the BC Biotech Association provides networking opportunities for its members — 80 percent of those interviewed mentioned the importance of the BC Biotech Association. BC Biotech is a non-governmental, not-for-profit, industry driven association. Representing over 90 percent of BC biotech's community across all sectors, the association provides a coherent voice on a range of issues that affect the industry, from influencing public policy to stimulating investor interest. It gives its members access to the information, resources and assistance they need to make their ventures a success. BC Biotech keeps its members informed about the current state of the biotechnology industry. Through its events, seminars, and educational opportunities, it facilitates networking, partnering opportunities, and growth for its members and for BC's biotechnology industry.

BC Biotech appears to be the glue that keeps these companies together. The networking events are more for social purposes, to know the community (for newcomers), and as a key for consultants and lawyers for making contacts. While BC Biotech is the glue, the driving forces are the research programs at UBC and the existence of venture capital and angel investors. Arguably there is little horizontal integration, but even less vertical integration, as in the Porter model of clusters.

Firms are anchored to the region by multiple factors. A connection to UBC seems to be a common anchor for many of the companies. Many of them started around the same time and went through similar challenges — having other individuals to interact with who have had similar experiences has created a shared bond between some of the firms. Firms rarely compete with each other for local talent. They are able to attract talent to Vancouver, aided by the weather, scenery and lifestyle of the city, and out-migration of talent does not appear to be a problem. Firms appear to be sufficiently specialized that there is little movement among firms by the technical specialists, thus creating a horizontal group of firms, each more or less at the same level of relationship to their sources of intellectual property (IP) and at the same level of relationship to their other inputs (capital, services) and their eventual clients/customers.

As a side experiment we searched Web sites for the biographies of the senior officers of most of the biotech firms in Vancouver. A majority of the sites listed the degrees, and the institutions that awarded them, for their senior people. While this experiment was not rigorous, 30 out of 44 sites examined listed the degrees for at least some of their key people. Forty-five percent of these people had degrees from Canada, with 27 percent coming from BC. About 30 percent had their highest degree from a US institution. One senior official of a biotech firm in Vancouver told us that most of their recruiting was done outside BC, and that if they could get their senior people to stay in Vancouver for two years, then they had little worry of them leaving. The first two years were critical; this manager emphasized the need for some sort of special tax treatment for new hires, particularly those from the US, to ease the transition from the US tax system to the Canadian system.

Another important element in BC is the existence of a strong venture capital market; several interviewees mentioned the benefit of having a number of angel investors in BC with deep pockets. Venture capitalists stated,

> "There are lots of venture capitalists in BC but not enough science."
>
> "The environment of the biotech cluster in Vancouver is similar to that in the mining industry in Vancouver when the Vancouver Stock Exchange was still operating."
>
> "Vancouver is used to high risk with unpredictable outcomes. There are a lot of angel investors because of this dynamic."

An analogy serves to illustrate the cluster: consider a garden in Vancouver (with the backdrop of mountains and sea!). The soil is the university environment, predominantly UBC. The seeds of ideas germinate in the soil there, watered by funding from the federal granting councils. The plants grow, fertilized by funding from venture capitalists. There is some cross-pollination from BC Biotech and more water from federal R&D support to companies (Industrial Research Assistance Program funding and R&D tax credits). Few of the flowers live long enough to mature, go to seed, and generate new plants; most are picked while they are still blooming and are carried off by passersby —

large multinational pharmaceutical companies, who want to have the flowers in their homes.[5]

OBSERVATIONS AND ANALYSIS

Over the past decade a stream of new companies has been spun off from one or other of the research facilities. This trend stopped two years ago, perhaps because of the downturn in the stock market and the parallel decline of dot-com companies. But this may not be the only reason. One respondent is quoted as saying that "There are a lot of genetic disease spinoffs that are making good money but there isn't enough science coming out of UBC, so there is currently a slow decline of spinoff companies."

As seen in Figure 1, the rate of biotech start-ups over the past decade has varied widely. There are no obvious reasons for this time variance: from 1992 until 2001 the stock market's appetite for initial public offerings (IPOs) was insatiable. What causes the flow of intellectual property from the universities (particularly UBC) to vary so greatly?

Are the sudden changes in the number of spinoffs due to market conditions or to other factors? It is probable that the decline of spinoff companies from UBC since the mid-1990s can be attributed to the following:

- not enough money to support new companies (high-tech bubble burst);
- saturation of the market (too many one-product/technology companies); and
- venture capitalists concentrating their money in already existing companies, trying to keep them alive.

Companies spun off from university research labs appear to be a particularly effective means of technology transfer out of universities (Rogers, Takegami and Yin 2003).

The importance of these spinoffs in the cluster leads to another question: What explains the difference in spinoff formation between UBC and SFU? UBC has a medical school and SFU does not, which for this sector is a major element as a driver and user of innovations. This division is emphasized by the tight concentrations of UBC spinoff firms around two UBC locations, its main campus and the teaching hospital. This feature is hardly surprising, but what is interesting is the much wider dispersion of the SFU spinoffs. But perhaps this is due to their radically different intellectual property policies. UBC retains all rights to discoveries made in its labs, granting only a non-exclusive licence to the discoverer, while SFU gives all intellectual property rights to the discoverer in its labs. In the case of SFU, the University Industry Liaison Office (UILO) acts only as a broker if the individual so desires.

Could UILOs have done better trying to merge individual initiatives and make stronger companies with more than one product in the pipeline? This appears to be a disadvantage for the growth of the companies and the cluster.[6] Of course, part of the problem is that each scientist wants his own company.[7] Possibly firms are spun off too early in the stage of technological development. They can have promising technology/products, but it is difficult to survive in the long term, needing ongoing injection of financial resources without producing results and revenues.

As noted above, there is also the interesting question as to whether Victoria is part of Vancouver's biotechnology cluster. No matter what method of transportation is used, it takes over two hours to travel from one city centre to the other (other than by scheduled helicopter service). Two hours is often taken as the outer boundary for travel time across the geographic area of a cluster. While some argue that information technologies permit the creation of virtual clusters, there are those who argue that innovation is primarily a matter of interpersonal communication (as for example, Antonelli 2000). Accepting this argument, it is clear that electronic means of communication (even Internet video) have not yet reached the level of acceptance where they can replace physical meetings.

What is clear is that there are many region-specific factors supporting the success of the biotech cluster in Vancouver. Vancouver has world-class university research facilities, but no large, multinational, private-sector enterprise. This may have region-specific results in terms of the creation and development of start-up firms. These firms are not dominated (or discouraged) by the presence of a large player in the local cluster, nor are they fed "scraps" of intellectual property from projects that a larger firm might not wish to follow up. Vancouver has always been home to a highly entrepreneurial, risk-taking financial community, originally established to exploit investment opportunities in the resources sector — mainly mining, but the resources sector has been all but taken over by large multinational firms. There is an observed abundance of entrepreneurial, risk-taking managerial and financial talent, and like the technical specialists these individuals have elected to transfer their entrepreneurial and intermediary skills to a new sector, rather than move elsewhere.

Also, the chartered banks, which do play a role in generating capital for the industry, have S&T investment officers with independent decision-making authority in Vancouver, unlike other regional centres such as Halifax. This may be an important factor, and it illustrates that national innovation policy should also focus on the established financial services sector, rather than simply assuming that all capital needs will be met through independent venture capitalists.

How then does the Vancouver cluster sustain itself and remain competitive at the national level, if not globally? The answer seems to lie in the nature of the outputs of the firms — they often do not manufacture and market their own products, but rather sell the intellectual property (including regulatory approvals and licences) to larger multinational firms for manufacture elsewhere. The Vancouver entrepreneurial environment seems particularly favourable to the creation of firms, and it is the firms themselves (or their major assets, usually intellectual property) that are the final product of the cluster.

By comparison, preliminary results suggest that the other two large centres of biotech activity in Canada, Montreal (Quebec), and Toronto (Ontario) have quite different characteristics, a result predicted by Table 2. Montreal is home to manufacturing and research facilities of several large multinational firms, induced — in part by the active promotion of Quebec as a research hub by the provincial government — through the use of preferential R&D tax credits. Toronto also has a large biotech manufacturing sector, but these firms for the most part are generic drug manufacturers that manufacture and market pharmaceuticals and other biotech based products whose patents have expired or whose properties are already well known. What all three centres have in common is a large, world-class university research community and a cosmopolitan location that provides support for all cultures and lifestyles and good transportation links.

CONCLUSIONS

This study is a work in progress, which will be extended to include the characteristics of other clusters being studied by other investigators, both those studying the biotech sector in other cities in Canada and other industrial clusters in Vancouver and elsewhere. The commonalities and differences noted above may be specific to the biotech industry in Vancouver. But what is clear is that these similarities and differences do have to be understood and applied to other centres in Canada. Governments, both federal and provincial, may wish to establish biotech clusters elsewhere in Canada to make use of local sources of highly skilled labour, raw materials or manufacturing capabilities. But simply wanting to have a biotech cluster establish itself, or survive in the long run without massive, ongoing injections of financial resources into such communities is not enough. There are necessary and sufficient conditions for the establishment of any industrial cluster in a community, and these conditions probably differ from one industrial sector to another.

The major policy issue that underlies this investigation is: What are the necessary and sufficient conditions that support the formation of a biotech cluster in Canada? Are the necessary or sufficient conditions region-specific?

What are the necessary (common) features: university, labs, government agencies, private firms, human capital? What are the sufficient conditions (conditions for continued existence)? According to Porter, they must include at least one private firm with a global reach. We propose a new test: Can the cluster survive the catastrophic loss of a node/actor, such as the closing or transfer of a major industrial facility? Can a cluster survive without certain nodes, whether they be public-sector or private-sector nodes?

The results suggest that innovation policy must not only focus on public investment in science and technology but also on issues such as venture capital financing, human capital development, and the factors that influence the quality of life in a city.[8]

The data to date suggest that there is indeed a viable biotech cluster in the Lower Mainland region of BC. There is a strong public-sector institution (UBC) at the centre of the cluster, supported by smaller, more specialized research institutes funded by both the federal and provincial governments. There is a continuous stream of highly qualified researchers coming through the postsecondary education pipeline into the cluster. There are a number of viable enterprises, with sufficiently diverse interests and markets that if one enterprise fails, the viability of the cluster is not immediately put in question.[9] The cluster is constantly evolving, changing as the technology evolves, and constantly seeking new market niches.

Unlike the classical definition of a cluster, the biotech cluster is not based on the manufacture and marketing of specific, physical products, but rather on the development of intellectual property. The intellectual property is often first created in one of the public-sector institutions, before it is transferred to the private sector through the licensing of the base technologies to a start-up company. Indeed, one can think of the public sector as acting as a catalyst for the creation of spinoffs.[10] These companies develop the intellectual property and bring it to the level where it is ready for production. At this point, the BC strategy is to sell its intellectual property, or the company as a whole, to an established biotech manufacturer — often a multinational pharmaceutical company.

This process does not fit the traditional cluster model constructed by Michael Porter. The Porter model has two features that narrow the scope of the concept to a large manufacturing-based economy with domestically-based multinational companies. Porter's model defines a cluster to be a vertically-integrated agglomeration of enterprises that have a strong domestic market and a significant competitive advantage in the global market.

> It is recognized that all user-producer (or customer-supplier) relations constituting inter-firm networks must, by definition, involve some degree of vertical integration. Therefore, horizontal is used to describe networks based on other kinds of relations — including, for example, cooperation between rivals or informal know-how trading. This usage has generated a basic distinction between "trade" networks with strong vertical aspect and "knowledge" networks with a strong horizontal aspect (Edquist and Hommen 1999, p. 73).

In general, the BC biotech sector does not manufacture commercial products — its product, if it has one, is intellectual property itself. As a corollary, the cluster is not a vertically-integrated agglomeration, but a loose horizontal association of enterprises that do not compete for market share. We have found this "knowledge network" model to be a viable one. Given that the BC economy is in a transition from a resource-based economy to a knowledge-based service economy, a horizontal model may be the successful model for this specific set of economic circumstances.

The rise of contract research organizations (CROs) is a case in point. Canada is a good place for a corporation to carry out research: professional salaries are on average lower than in the US, and there is a favourable tax regime for corporate R&D at both the federal and provincial level. The CROs have few local customers; they make use of the existence of the Vancouver biotech cluster to develop their capabilities and export their knowledge-based services.

Suh (2002), in his description of emerging patterns of innovation networks in Korea linked the emergence of "new technology-based firms," mainly small and medium-sized enterprises (SMEs), to the economic upheavals of

1997, when the traditional Korean economic system founded on the *chaebols* was shattered. It is tempting to link the emergence of a regional cluster based on SMEs to the disruption, in a Schumpeterian sense, of the existing economic order in a region, leading to the creation of conditions where radically new enterprises can flourish. This is certainly true for BC: the severe contraction of its resource-based economy due to depressed world commodity prices, trade disputes with the US, natural disasters (forest fires) and high labour wage rates have certainly provided an economic disruption at the regional level, equivalent to the Asian financial crisis of 1997.

There are several high wage rate, high educational-attainment economies based on resource extraction. These economies are constantly being threatened by competition from lower-wage rate (and usually lower educational-attainment) resource-based economies. It is tempting to suggest that it is possible for an economy to evolve from being a resource-based economy to a knowledge/service-based economy without having to pass through the intermediate stage of being an industrialized manufacturing-based economy. The BC biotech cluster offers, at least, a potential blueprint for this type of transition. Study of this example may well give policymakers insight into the conditions that they may wish to replicate in other jurisdictions and for other industrial clusters. Cluster size is important — there are probably critical factors, below which cluster activity will not ignite and be self-sustaining, such as population, regional domestic product, access to human resources from outside the region, transportation, and communications infrastructure.

NOTES

[1]Two key differences are (i) the cluster does not have to be globally competitive, but nationally competitive, and (ii) public sector institutions can (and do) play major roles in the formation and sustainability of the cluster. Rees (1999) argued there was no biotech cluster in Vancouver in 1997 in the classic Porterian sense.

[2]BC Biotech (2002) stated there are 88 firms in Vancouver and the southern part of Vancouver Island.

[3]As noted by Richard Florida (2000), people often move not for the job that is being offered to them, but for the job opportunities presented by moving to the new location.

[4]The social benefits conferred by the higher tax system in Canada include health care. As well, the relative advantage in purchasing power of the Canadian dollar has erased the perceived differential. Also from a firm's point of view costs per researcher

are lower: R&D expenditures per researcher (in US$ in 1999, as reported by the OECD) are $112,000 per researcher in Canada, $168,000 in France, $148,000 in Germany, $135,000 in the UK, and $160,000 in the US.

[5]We are indebted to Monica Salazar for developing this analogy. It is left to the reader to determine the nature of the weeds and other pests usually found in my garden!

[6]This idea was mentioned by several interviewees, especially the consultants and venture capitalists.

[7]The "his" is used on purpose, since there are few female entrepreneurs. Julia Levy of QLT is one.

[8]The reader should review several of the papers by Richard Florida on this subject; his results, which were based on research in the United States, have been replicated for Canada by Gertler and others and can be found at <www.competeprosper.ca>.

[9]This is a feature of the horizontal nature of the cluster — the firms are not interdependent and act in different technology/market niches.

[10]The Canadian R&D tax credit program is also a major public-sector support mechanism but it is not region-specific, although some provinces "sweeten" the benefits in order to try to attract R&D enterprises. The tax credit/cash rebate provisions of this program also set in place a significant barrier to small privately-controlled firms who are considering the transition to publicly traded status.

REFERENCES

Antonelli, C. 2000. "Collective Knowledge Communication and Innovation: The Evidence of Technological Districts," *Regional Studies* 34(6):535-47.

Asheim, B. and A. Isaksen. 2002. "Regional Innovation Systems: The Integration of Local Sticky and Global Ubiquitous Knowledge," *Journal of Technology Transfer* 27(1):77-86.

BC Biotech. 2002. "Vancouver: A North American Biotechnology Centre," *Report.* At <www.bcbiotech.ca>.

Cooke, P. 1998. "Introduction: Origins of the Concept," in *Regional Innovation Systems: The Role of Governances in a Globalized World*, ed. H.-J. Braczyk, P. Cooke and M. Heidenreich. London: UCL Press.

Edquist, C. and L. Hommen. 1999. "Systems of Innovation: Theory and Policy for the Demand Side," *Technology in Society* 21:63-79.

Florida, R. 2000. "The Economic Geography of Talent." Carnegie Mellon University. Unpublished manuscript. At <www.utoronto.ca/isrn>.

—— 2002. *The Rise of the Creative Class: And How It's Transforming Work, Leisure, Community and Everyday Life*. New York: Basic Books.

Goldfarb, B. and M. Henrekson. 2003. "Bottom-Up versus Top-Down Policies Towards the Commercialization of University Intellectual Property," *Research Policy* 32:639-58.

Holbrook, J.A. and D.A. Wolfe. 2000. "Introduction: Innovation Studies in a Regional Perspective," in *Innovation, Institutions and Territory: Regional Innovation Systems in Canada*, ed. J.A. Holbrook and D.A. Wolfe. Kingston and Montreal: School of Policy Studies, Queen's University and McGill-Queen's University Press.

Longhi, C. 1998. "Networks, Collective Learning and Technology Development in Innovative High-Technology Regions: The Case of Sophia-Antipolis," *Regional Studies* 33(4):333-42.

OECD. 1997. *National Innovation Systems.* Paris: OECD.

—— 2001. "Provisional Single Definition of Biotechnology," 2nd ad hoc meeting on biotechnology studies, DSTI, OECD, Paris, 3-4 May 2001.

—— 2002. *Territorial Review of Canada.* Paris: OECD.

Porter, M. 1998. "Cluster and the New Economics of Competition," *Harvard Business Review* (November-December):77-90.

Queenton, J and J. Niosi. 2003. "Bioscientists and Biotechnology: A Canadian Study." Presentation at the Third European Meeting on Applied Evolutionary Economics, Augsburg, Germany. At <www.emaee.net>.

Rees, K. 1999. "Innovation in the Periphery: Networks or Fragments in the High Technology Industry of Greater Vancouver," PhD Thesis, Department of Geography, Simon Fraser University, Burnaby.

Rogers, E.M. 1983. *Diffusion of Innovations.* New York: The Free Press.

Rogers, E.M., S. Takegami and J. Yin. 2003. "Lesson Learned about Technology Transfer," in *Systems and Policies for the Globalized Learning Economy,* ed. P. Conçeicao, D.V. Gibson, M.V. Heitor and C. Stolp. Westport: Greenwood Publishing Group.

Suh, J.-H. 2002. "The Emerging Patterns of SMEs Innovation Networks in Korea." Report for the World Bank project on *Restructuring of SMEs for the Knowledge Economy — Role of Public Policy in Korea.* Seoul: Korea Development Institute.

Wolfe, D.A. and M.S. Gertler. 2003. "Clusters Old and New: Lessons from the ISRN Study of Cluster Development," in *Clusters Old and New: The Transition to a Knowledge Economy in Canada's Regions*, ed. D.A. Wolfe. Kingston and Montreal: School of Policy Studies, Queen's University and McGill-Queen's University Press.

6

NETWORKING AND INNOVATION IN THE QUEBEC OPTICS/PHOTONICS CLUSTER

Mélanie Kéroack, Mathieu Ouimet and Réjean Landry

INTRODUCTION

It is increasingly the case that to stay competitive, firms must innovate. Although a firm's capacity to innovate depends primarily on internal capabilities and resources (Von Hippel 1988; Lundvall 1988; Hakansson 1989), interactions with clients, suppliers, research institutions, etc., can enhance this capacity by facilitating broader access to knowledge. Taken together these interactions can form networks that facilitate knowledge exchange among a range of institutions. Under certain conditions these networks may be manifest as geographically based clusters of organizations that create and sustain a regional competitive advantage.

Despite the many studies that describe the emergence and evolution of clusters, few deal with clusters in the optics/photonics industry. This chapter brings cluster analysis to bear on this important industry by investigating the interorganizational linkages between the actors in the Quebec optics industry. This study is based on a survey of 47 organizations involved in the optics/photonics industry (n=58) in Quebec City. The literature on clusters frequently notes the importance of interorganizational linkages and the social networks they form, but does not systematically examine the phenomena. In contrast, this chapter adopts a systematic sociometric approach to analyzing the relationships within the industry, focusing on the frequency and nature of interactions.

The remaining sections of this chapter are as follows: we first discuss the literature dealing with the cluster concept in general and optics/photonics clusters in particular; we then discuss methodology and data collection and present our findings; in the final section we offer our analysis and suggest implications for further research and public policy.

PRIOR STUDIES

Prior Studies on Clusters

The concept of clusters is part of a new understanding of regional and national economic development in which innovation is the core determinant of competitive advantage for the regions and the firms. The tremendous success of certain regions (Emilia Bolognia, Silicon Valley, etc.) has focused new attention on ways in which regions can foster innovative economies (Goodman *et al.* 1989; Brusco 1990 quoted in Hendry, Brown and Defillippi 2000). Much of the recent interest in industry clustering stems from the work of Michael Porter. Porter defines clusters as "a geographically proximate group of interconnected companies and associated institutions in a particular field, linked by commonalities and complementarities" (Porter 1998*a*, p. 199). The Organisation for Economic Co-operation and Development (OECD) defines clusters as a "network of production of strongly interdependent firms (including specialized suppliers), knowledge producing agents (universities, research institutes, engineering companies), bridging institutions (brokers, consultants) and customers, linked to each other in a value adding production chain" (OECD 1997, p. 5).

The literature on clusters has grown rapidly among both academics and policymakers. Many researchers have applied alternative concepts such as industrial districts, innovative milieux, or innovation systems to capture the regional dynamics of economic development and innovation (Braczyk, Cooke and Heidenreich 1998; de la Mothe and Paquet 1998; Enright 2000; Holbrook and Wolfe 2000; Landry and Amara 1998, 2001; Lawson 1997; Le Bas *et al.* 1998; Liyanage 1995; Lundvall 1988; Maillat 1995; Nauwelaers and Wintjes 2001; Porter 1998*b*, 2000; Putnam 1993; Rosenfeld 2002*a*,*b*; Storper1997; Turner 2001). The central element of the cluster literature is the hypothesis that clusters enhance a firm and region's competitiveness by facilitating the creation and diffusion of knowledge (Malmberg and Power 2003). Today, a firm's survival depends heavily on advantages generated by the creation and absorption of knowledge. According to the OECD, "today, knowledge in all its forms plays a crucial role in economic processes. Nations which develop and manage effectively their knowledge assets perform better. Firms with more knowledge systematically outperform those with less" (OECD 1997, p. 15). Hence, firms are participating in formal and informal networks to enhance their ability to learn through increased interactions with other actors. Some scholars argue that the geographical proximity inherent in clustering enhances innovation by facilitating "learning-by-interacting" (Dicken *et al.* 1994; Saxenian 1994). Sociologists have long asserted the importance of spatial proximity to facilitate the creation of relationships (Romijn and Albaladejo 2002). In economic development Marshall (1920) was the first to emphasize the importance of proximity by using the expression "industrial atmosphere." Some studies demonstrate that the frequency of contact and communication between actors increases with geographic proximity and in turn enhances the diffusion of innovations within a region (Darr and Kurtzberg 2000). The main advantages generated by proximity are the reduction of uncertainty and the increase of tacit knowledge exchange. It is now acknowledged that geographical proximity has an important role to play in the competitive success of firms and more precisely in network development (Porter and Stern 2001).

In short, the concept of clusters highlights the importance of interorganizational links in facilitating the development of complementary capabilities and resources and the exchange of knowledge (Landry and Gauthier 2003, p. 8). Henceforth, the object of analysis in regional economic development is not the firm itself, but the network of ties between firms, in particular the intensity and diversity of interactions between firms and other actors in a cluster (clients, suppliers, competitors, government, university, research centres, support organizations, etc.). The next section will provide a general view of the previous literature specifically related to optics/photonics clusters.

Prior Studies on Optics/Photonics Clusters

Optics, also known as opto-electronic, photonics, and optical science and engineering, is an important and fast growing industry with applications in a large variety of markets. Optics/photonics technologies involve the production, manipulation, transmission, and detection of photons, fundamental components of light composed of waves and energy particles.[1] Until recently the information and telecommunications industries were the largest markets for optical technologies, but optical applications are being introduced in other sectors as well: aerospace, automobile, biomedical, defence, environment, forestry, industrial process, safety, transport, retailing, etc.

One can divide this industry into three categories: generic technologies, key components, and end-user products and systems (Miyazaki 1995). Many companies are active in all three product lines. For example, Nortel uses generic technology to manufacture its own components, which it inserts into its telecommunications systems that will later be sold to service providers (end-user products) such as BELL and AT&T.

The modern applications of optics occurred after World War II. The period from the 1960s to the 1970s was mainly dedicated to research and development, and the 1980s and 1990s to the diffusion of new technologies (Miyazaki 1995). Initially, interest in optic products was mostly among large firms. During the early 1990s entrepreneurial personnel from large companies, government research institutes, and universities began leaving their organizations to create new firms.

TABLE 1

Optics Industry Categories and Descriptions

Category	Examples
Generic technologies and materials	• Optical glass • Epitaxial wafers • III-V materials • II-VI materials • Glass fibre
Key components	• Lasers • Fibre optic cables • Light emitting diodes • Sensors • Optical assemblies
End-user products and systems	• Telecommunications • Night vision systems • Optical storage • Process control • Image systems

Source: Miyazaki's three-level model of the opto-electronics industry, with examples of typical products at each level (Hendry *et al.* 2001).

Of the few studies on optics/photonics clusters, most simply describe the cluster's development, composition, and functioning and make little attempt to understand the linkages that sustain it (Catts 1999; Empire State Development Division of Policy and Research 2001; Landry and Gauthier 2002; Mallett 2002; USF office of Economic Development 1999). Chris Hendry (Hendry, Brown and Defillippi 2000) is one of the few researchers who have carried out studies directly related to optics/photonics clusters. One of his most interesting studies investigates the extent and significance of localized inter-firm trading and network relationships in six optics/photonics clusters located in Wales and East Anglia (UK), Arizona and Massachusetts (US), and Bavaria (primarily Munich) and East Thuringia (Germany). The main finding of Hendry's work is that national and international relationships are found to be much stronger than local ones in the optics/photonics sector. Hendry argues that as firms move away from short-run production to production processes that have a greater scientific and technological content, the local supply base of labour, technology and knowledge diminishes and firms are forced to seek commercial connections in a much wider geographic and technological space. This conclusion agrees with the evidence other researchers have found contradicting the importance of the proximity benefits paradox (Suarez-Villa and Walrid 1997; Sternberg 1999; Simmie 1997; Larsson 1998, quoted by Romijn and Albaladejo 2002). These authors suggest that interactions with international partners are more important than proximity to customers, suppliers, and competitors to develop or improve products and manufacturing processes (Amara and Landry 2003, p. 16).

As mentioned by Hendry (Hendry, Brown and Defillippi 2000), studies of optics/photonics clusters are conspicuously lacking and more are needed to help us better understand their particular characteristics and dynamics.

METHODOLOGY

Qualitative Data

The population of this study was identified through the Québec – Optics City's official directory, which includes a list of 58 actors. These actors were divided into six categories: 22 firms, five research institutes, six educational institutions, six government organizations, seven financial institutions, and 12 local development support organizations. The chief executive officer (CEO) for each organization was identified and sent a letter explaining the project. All the participants were contacted by phone in order to schedule an appointment. The response rate for the study was as follows: 18 out of 22 firms, four out of five research institutes, four out of six educational institutions, five out of six government organizations, five out of seven financial institutions, and 11 out of 12 local development support organizations. Overall, we had a participation rate of 81 percent. A total of 47 face-to-face interviews were held from January to May 2003. All the interviews were recorded and transcribed. The average length of an interview was approximately one hour.

The interviews were conducted with the Innovations Systems Research Network (ISRN) questionnaire, which is based on the cluster literature and the Oslo Manual methodological guidelines (OECD 1997). The main topics covered by the questionnaire were the company's background, research and innovation strategy, networking relationships (suppliers, customers, competitors), locational infrastructure factors, role of research institutes and technology transfer centres, local cluster characteristics (social capital), and the future of the cluster (cluster's strengths and weaknesses). The questionnaire was suitably modified for use with government organizations and research institutes. Qualitative and quantitative methods were used during the interviews. At the end of the interviews, respondents were asked to measure the degree of novelty of the innovation, to identify the main sources of new employees, the factors contributing to the emergence of the industry in Quebec, and to measure the interorganizational linkages within the industry.

Network Data

Network data describe the relational ties linking different actors and were collected via a short sociometric questionnaire that listed all the actors of the Quebec optics industry. The boundary, the list's content, of the cluster was primarily selected by following a positional approach (Scott 2000, pp. 55-58), which is done by selecting all actors that figured in the directory of the Québec – Optics City. In order to prevent omissions, we also used a reputational approach by asking participants to nominate other actors who were not included in the directory. New nominees were then contacted for an interview. At the end of the data-collection process, a questionnaire with just a list of the new nominees was sent back to the participants. This process allowed us to gather network data on the industry's entire population.

The sociometric questionnaire included the following two questions: How frequently does your organization have contact with the following organizations (1=never, 2=rarely, 3=sometimes, 4=often and 5=very often)? During these contacts, did you discuss: market development, financing, training, production processes, research and development, services provided by support organizations, market situation, and/or other, specify? Thus, the first question allows us to measure the strength of each tie. According to Granovetter "the strength of a tie is a (probably linear) combination of the amount of time, the emotional intensity, the intimacy (mutual confiding), and the reciprocal services which characterize the tie" (1973, p. 1361). In this study, tie strength was measured by using frequency criteria. Strong ties refer to the "often" and "very often" responses and weak ties to the "rarely" and "sometimes" responses. Our operationalization of strong ties assumes that frequent or very frequent interactions take more time and generate more emotional intensity and more intimacy than less frequent ties. The second question allows us to identify the discussion topics that are linked with every tie. It must be noted that his chapter only analyzes data collected from the first question. Data collected with the second question will be used in forthcoming papers.

Approach to Network Data Analysis

We used the linkage data to look for structural properties at both the network and the actor levels. At the network level we are primarily interested in determining the density of the cluster by estimating the proportion of strong versus weak ties in the cluster as a whole and for all sub-networks, such as the relations between firms, between non-firms, and the relations between firms and non-firms. At the organizational level our objective is to determine the centrality of various actors by identifying the cluster's most connected and most intermediary organizations.

To compute the network-centred and organization-centred measures, we built various squared "actors by actors" matrices by using the UCINET software (Borgatti, Everett and Freeman 2002). The calculation of density and centrality required that data matrices be symmetric. A symmetric "actors by actors" matrix is one in which all ties are reciprocal. An example of symmetry is when the respondent of organization A declares that his organization enters frequently into contact with organization B and the respondent of organization

B confirms that his organization enters frequently into contact with organization A. The fact that we are dealing with organizations that include many individuals, or potential respondents, may explain why the condition of symmetry rarely holds. The transformation of asymmetric data into symmetric data was performed by using the following approach: for each pair of answers, we took the highest frequency score (*never*, *rarely*, *sometimes*, *often* or *very often*), except for the relations between firms and non-firms, for which we accepted the firm's response. We hypothesized that non-firms would tend to overestimate the frequency of interactions that they have with firms because of their mission. Finally, organizations that did not participate in our study were included in the analysis when participating firms mentioned interacting with them. In these cases we accepted the answers given by the participating agencies without qualification. Keeping non-participating organizations in the analysis allows for the production of more realistic results. The only ties that are missed are the ones between non-participating agencies.

The network analyses presented in this study called for the construction of 12 matrices. First, we constructed matrices covering both weak and strong relations (weak and strong): one included all organizations within and related to the optics industry, one included firms only, one included non-firms only, and one included relations between firms and non-firms only. We then repeated this exercise for weak and strong relations independently. For the matrices of both weak and strong ties the answers never were coded 0, while answers rarely, sometimes, often and very often were coded 1. For the matrices including only the weak ties, the answers rarely and sometimes were coded 1, while other answers (never, often and very often) were coded 0. For the matrices including only the strong ties, the answers often and very often were coded 1, while other answers (never, rarely and sometimes) were coded 0. These 12 matrices allowed us to compute the proportion of weak versus strong ties in the optics industry as a whole as well as in three sub-networks (relations between firms, relations between non-firms, and relations between firms and non-firms). The centrality scores of each organization were computed by using the matrix including weak and strong relations within the whole industry. The results of the network analysis are presented in the section titled "Cluster Capacity to Exchange Knowledge."

FINDINGS

Quebec Optics Industry Initial Steps

To better understand why the optics/photonics industry has emerged in Quebec City, we now examine some of the most significant events that preceded it. The discovery of the TEA CO2 Laser in 1960 by Defence Research & Development Canada (DRDC) located in Quebec City completely revolutionized the optical industry around the world. This discovery launched numerous laser applications (fiber optics for communications, laser surgery, compact discs, etc.). Many of our respondents claimed that this discovery significantly increased the interest of industry, universities, and governments in the optical field and was the precursor to the development of an optic industry in Quebec City. Optical applications in the defence area are still important. Defence expenditures have, over time, been the greatest economic driver of advanced optics technologies in Quebec and almost every weapons system reflects that investment (Catts 1999, p. 8).

Likewise, the discovery of the TEA CO2 promoted a new interest in optics among researchers at Laval University. In 1964, a group of researchers from the physics department began to conduct research and train students on optics.[2] This group also started to collaborate with private sector partners early on. In 1989, they formally established The Centre for Optics, Photonics and Laser (COPL), which came to occupy a leadership position in research. COPL quickly became well-known for the high quality of its students. Furthermore, two community colleges, Le Cégep de la Pocatière and Le Cégep de Limoilou also initiated technical training programs in optics for the emerging industry in early 2001.

In the 1980s, the Quebec City area was home to about half of Canada's researchers in optics and photonics. The present number of academic research institutes in optics was reached in 1985 with the establishment in Quebec City of the Institut National d'Optique (INO).[3] The activities of COPL and INO greatly facilitated the transition from research to industrial applications. Some of today's largest firms in the industry were created in the 1980s: Exfo, ABB Bomem, and Gentec. The Quebec optics industry is a good example of a science-based industry that has been incubated in quasi-government and academic laboratories for a long time before an industry gradually grew up around it through incubation of new firms.

Governance

The actors in the Quebec optics/photonics industry were formally introduced to the cluster concept in 1998 during meetings with Bob Breault, a prominent figure in the creation of the optics cluster in Tucson, Arizona. These discussions were instrumental in the creation of the Quebec Optics and Photonics Association (GOPQ), the Quebec clustering organization dedicated to the promotion of the industry in the region and abroad.

In 1999, a second clustering organization, the Québec – Optics City (Québec – Cité de l'optique), was established to support the development of the optics and photonics industry in the Quebec City area. The main goal of Québec – Optics City is to support existing companies, assist new start-ups, and attract companies specializing in optics from outside the region. The establishment of the business plan for this newly created organization provided the first opportunity for all the actors in the region (companies, research centres, support organizations, etc.) to sit at the same table to work on a common project. Some people describe this event as the industry's first real clustering experience because Québec – Optics City had more resources and a stronger institutional base than the GOPQ.

In May 2003, during the completion of our data-collection phase, Québec – Optics City and GOPQ were agglomerated in a new regional economic development organization, Pôle Québec Chaudière-Appalaches, an umbrella organization overseeing the promotion of five industry clusters. This consolidation culminated in a consultation process which led the business community to assume that there were too many small organizations promoting too many industries in an uncoordinated strategic mode. The mission of this new organization is to catalyze the economic development efforts of the manufacturing and science-based industries, particularly those industries, especially optics and photonics, for which the region has distinctive competitive advantages. Pôle Québec Chaudière-Appalaches is rapidly emerging as the most important forum where the business community and the other actors supporting the regional economic development meet to foster a common strategy for promoting key clusters in the region. The Pôle Québec Chaudière-Appalaches has credibility because it has enlisted a number of high-profile members from the region's most vibrant industries to build a cluster development strategy.

Industry's Characteristics

The Quebec optics industry and supporting institutions are primarily composed of 22 firms, three major research centres, three training centres, four venture capital firms and ten support organizations.[4] The number of firms has increased from nine in 1999. Sales increased from $150 million in 1999 to $300 million in 2001. They have since decreased to $214 million in 2003. A similar trend can be observed in the R&D expenditures, which grew from $17 million in 1999 to $31 million in 2001 and then fell to $21 million in 2003. The firms in the industry are mostly small and medium-sized enterprises (SMEs). The average number of employees per firm is 56 with 64 percent having less than 50 employees. Overall, the industry increased its employees from 769 in 1999 to a ceiling of 2,095 in 2000. Employment stands at 1,548 in 2003. Key firms in the industry are EXFO (ranked 66th among Canadian firms in R&D expenditures), ABB Bomem, and Gentec. The firms show a high level of diversification. They manufacture products dedicated to many different market niches, notably telecommunications, biomedical, retailing, pharmaceutical, and the wood industry. Table 2 presents examples of the industry's product range. The firms are export-oriented. The majority of them make more than 80 percent of their sales abroad and are usually small players compared to their international competitors. This high level of export suggests that the firms benefit from ideas, information, and knowledge from clients located outside the region. Such openness to the world is a necessary condition to ensure the long-term development of clusters (Cowan and Jonard 2003).

To better understand the attributes of these firms it is useful to compare the Quebec City industry with other clusters. We have done so by using data gathered by Hendry (Hendry *et al.* 2001). As we can see from Table 3, the firms located in the Quebec City region tend to be smaller and younger than the firms operating in the other optics clusters (except for the Thuringia cluster). The proportion of independent private firms is also much higher in Quebec. In terms of products, firms in Quebec do not produce generic products as is the case in other clusters. Contrary to the other clusters considered in Table 3, a higher proportion of Quebec firms have been spun off from research organizations rather than from other firms. Quebec firms also rely much more extensively on national investments than in the other clusters where foreign investments are much more significant. Finally, as can be seen in Table 3, the number of firms operating in the Quebec region is much smaller than in many other clusters.

TABLE 2
Examples of Optics/Photonics Products

Companies	*Product Description*
ABB Bomem	Analytical solutions (i.e., oil and fat analyses)
CorActive	Erbium doped optical fiber (i.e., boosting the transmission speed of information in telecommunication system)
Cybiocare	Medical device to improve diabetes control (i.e., watch with integrated glucometer)
Exfo	Optical testing and measurement (i.e., fiber installation and maintenance)
Fiso Technologies	Fiber optics sensors (i.e., optic in-vivo pressure transducer)
Infodev	Counting systems (i.e., counting people electronically (bus, mall, etc.))
InSpeck	3D full body digitizer (i.e., movies and plastic surgery)
Optel Vision	Machine vision inspection systems for high-speed packaging lines. (i.e., pharmaceutical industry)

The science-based firms of the Quebec region are supported by an impressive research infrastructure made up of three major organizations: the Institut National d'Optique (INO) with 240 researchers, the Centre d'optique, photonique et laser (COPL) at Laval University with 130 researchers and the Centre de recherche pour la défense of the Department of Defence at Valcartier with 350 researchers. The mission of these organizations is to support the development of applications for the private sector and each has a long track record of working productively with private partners. This research infrastructure has played a key role in the emergence of the industry and is likely to play an important role in ensuring the industry's capability to evolve.

The firms are also supported by a training infrastructure at the doctoral and postdoctoral levels at Laval University and other research organizations.

TABLE 3

Optics/Photonics Clusters Comparison

Dimensions	*Québec* Can*	*Wales UK*	*East Anglia UK*	*Arizona US*	*Masachusetts US*	*Bavaria EU*	*Thuringia EU*
			In percentage (%)				
Employees							
< 10	36	24	19	35	20	10	30
10–100	55	38	73	47	35	70	50
100–500	9	38	8	18	25	20	20
> 500	0	0	0	0	20	0	0
Years of operation in the region							
< 5	59	14	0	30	10	20	70
5–10	23	24	27	30	20	10	30
10–20	14	24	50	30	25	30	0
> 20	4	38	23	10	45	40	0
Product(s) main category							
Generic	–	33	16	4	17	6	0
Component	27	37	52	43	50	55	53
End-product	59	30	32	53	33	39	47
Ownership							
Independent private	77	44	46	76	35	50	40
Quoted	9	0	4	0	15	0	0
National subsidiary	14	22	12	12	40	20	60
Foreign subsidiary	–	34	38	12	10	30	0
Start-up pattern							
Compagny spin-off	22	57	61	29	50	70	80
University spin-off	22	5	23	41	35	0	20
Research center spin-off	28	–	–	–	–	–	–
Various spin-out	28	–	–	–	–	–	–
Investment							
National investment	94	29	8	18	5	0	0
Foreign investment	6	9	8	12	10	30	0
Sample							
Number of sample firms	18	18	23	19	20	10	10
Number of firms in region	22	25	53	125	370	152	–
Response rate	82	72	43	15	5	7	–

Note: The data in the first column have been added by us to facilitate comparisons.

Source: Hendry, Brown and Defillippi (2000).

Technical training is offered by two community colleges: the Collège de la Pocatière and the Collège de Limoilou. Laval University and the community colleges work in partnership with the clustering organizations to ensure that graduates have the skills necessary to meet the industry's needs. This training infrastructure represents another necessary condition to ensure the development and sustainability of a science-based cluster. However, during the interviews, many CEOs pointed out that it is sometimes difficult to attract highly qualified personnel capable of managing science-based firms and marketing science-based products in the region.

Only a few sources of capital are available to firms. A large amount of venture capital is available in the Quebec City area, especially Innovatech Québec-Chaudière-Appalaches, which is especially active in the optics/photonics industry. Firms have access to R&D tax incentives offered both by the Canadian and the Quebec governments. In 2003, the newly elected Liberal government in Quebec reduced the rates of many of the R&D tax incentives offered to science-based firms. Early evidence suggests that this is changing the climate of investment in science-based firms and reducing the capability of such firms to raise venture capital. The interviews also indicate that while start-up capital is abundant, the region suffers from a lack of venture capital firms with levels of capital required to ensure the development of firms. Overall, firms in the Quebec region have access to a large variety of sources of capital that are needed to ensure their long-term development.

To sum up, the firms operating in the Quebec region can rely on external supports in research, training, and venture capital. These organizations have long proven their ability to work collaboratively and productively with the optics firms. We will now turn our attention to the interactions and exchange of knowledge that exist between the firms and between firms and other actors operating in the industry.

INDUSTRY'S CAPACITY TO EXCHANGE KNOWLEDGE

Exchange with clients. Prior studies on innovation (Lundvall 1992; Von Hippel 1988) have noted the importance of clients as sources of knowledge to help a firm develop or improve products or processes. The cluster literature, especially that in the Porter tradition, claims that vibrant clusters require a large base of demanding, sophisticated local clients. This is not the case for the Quebec optics industry. These firms have very few clients located in the Quebec

City area. This agrees with other findings that optics/photonics clusters in general show a high level of international activities (Hendry and Brown 1998; Hendry, Brown and Defillippi 2000; Hendry *et al.* 2001). Two-thirds of the firms interviewed had less than 1 percent of their clients located in Quebec City and two-thirds had, on average, 95 percent of their clients outside the country. We should not conclude from these findings that the exchange of ideas, information, and knowledge with clients is made more difficult because there are very few sophisticated local clients. It must be pointed out that one of the optics/photonics industry's characteristics is the long sale-cycle. Furthermore, in the small niches occupied by Quebec firms, one can expect to have very few sophisticated local clients. To develop products, these firms have to work closely with potential clients to understand their needs and expectations. It is common for a firm to spend from 6 to 12 months working with a client. Since products have to achieve precise functional properties in conjunction with other components and systems, their production depends on close interaction, learning, and feedback between manufacturer and clients. Our findings clearly do not support Porter's hypothesis that a large base of sophisticated local clients is a necessary characteristic of a cluster.

Suppliers. As for the suppliers' localization, 63 percent of the firms buy more than 50 percent of their supplies in the Quebec City area. However, the interviews indicate that firms do not consider their local suppliers important sources of ideas, information, and knowledge to develop or improve products or manufacturing processes. The strong links, however, suggest that suppliers could play a significant role in fostering innovation in the industry. Further investigation will be needed to shed light regarding their distinctive contributions in the matter of product/process development.

Labour mobility. The innovation literature frequently claims that labour mobility is the major driver of knowledge flows between firms. Power and Lundmark (2003) found that the most successful industrial clusters in Stockholm were the ones with the highest rates of inter-firm labour mobility. Another recent study (Malmberg and Power 2003) arrives at similar results. Overall, the results of our interviews show that inter-firm labour mobility is significant: the Quebec firms hired 47 percent of their production personnel, 57 percent of their sales/marketing personnel, 31 percent of their scientists/engineers, and 57 percent of their managers from other firms in the region. Approximately 33 percent of the sales/marketing personnel were hired from outside Quebec City, compared to 14 percent of the managers, and 6 percent of the scientific and engineering personnel. Nearly 66 percent of the scientists and engineers have been hired from local research organizations or educational institutions. These findings suggest overall that knowledge flows are facilitated and fostered by these high rates of interorganization/firm labour mobility.

Competitors. The data collected in the interviews show that 7 percent of the direct competitors of the Quebec firms are located in the Quebec City area, 5 percent in the rest of Canada, 58 percent in the United States, and 30 percent elsewhere in the world. Overall, 81 percent of the firms interviewed indicated that the proximity to competitors did not influence the development and improvement of products and manufacturing processes. As noted earlier, the high degree of complexity and customization of the products developed in the Quebec industry requires sales representatives and engineers to spend a significant amount of time with their clients. Our interviews indicated that these interactions also provided opportunities to learn about their competitor's products. When asked about the global market shares of their products, most firms were unable to provide a precise answer. This lack of precise information about competitors is likely related to the fact that most Quebec firms occupy small, specialized niches where the number of clients and competitors is small and to the fact that Quebec firms may lack marketing expertise. Other studies on innovation show that small firms are less able than large firms to identify their competitors.

Cluster Capacity to Exchange Knowledge: The Sociometric Perspective

As mentioned in the previous section on research methodology network, data were gathered to identify and measure the strength and content of interorganizational relations. Figure 1 shows the graphical visualization of every relationship without making any distinction between weak and strong relations.

The visualization was drawn using the *Netdraw* software. The MDS (multidimensional scaling) procedure was used to position the more connected organizations at the centre and the less connected organizations at the periphery. Figure 1 shows that no organization is completely isolated. This means that every organization has at least one connection in the network. The graph's density is 0.48, which means that 48 percent of the possible relations are effective. When considering only the relations between firms, this proportion increases to 52 percent.

FIGURE 1
The Social Structure of the Quebec Optics/Photonics Industry

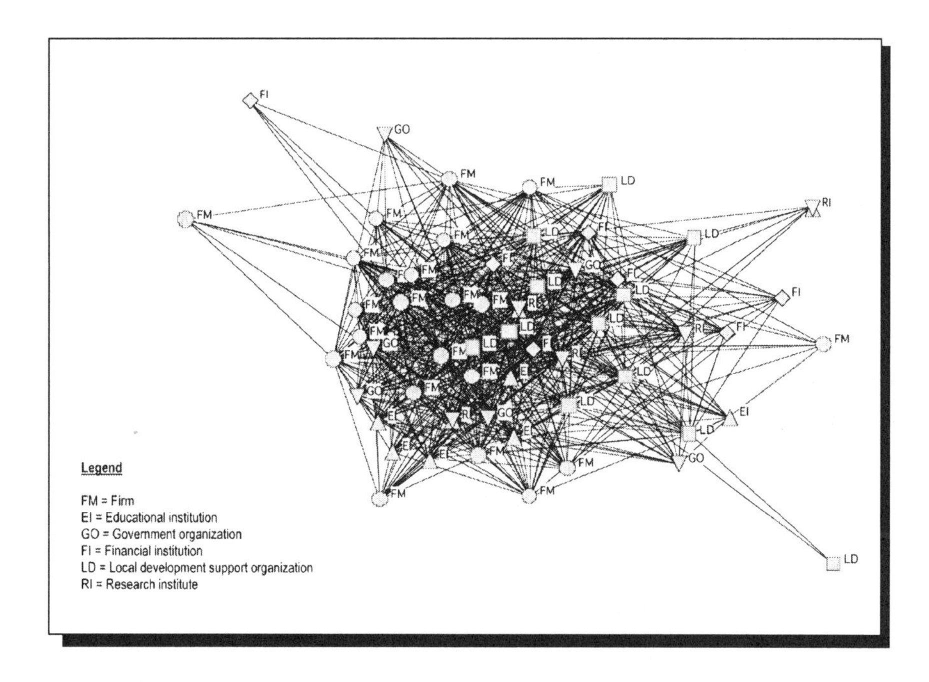

TABLE 4
Percentage of Strong and Weak Ties for the Cluster as a Whole and for Three Sub-networks

Ties	*Weak*	*Strong*
	In % of ties	
Within the whole cluster	62.6	37.4
Between firms	78.5	21.5
Between non-firms	55.3	44.7
Between firms and non-firms	63.5	36.5

Table 4 shows the percentage of strong and weak ties for the Quebec optics industry as a whole, for the relations between firms, for the relations between non-firms, and for the relations between firms and non-firms. We must recall that strong ties refer to the "often" and "very often" responses that respondents gave to the question: "How frequently does your organization have contact with the following organizations?" Inversely, weak ties refer to the "rarely" and "sometimes" responses. Almost two-thirds (62.6 percent) of the relations in the cluster are weak, while a little more than one-third of them are strong (37.4 percent). This proportion of strong versus weak ties is almost the same in the relations between firms and non-firms for which 63.5 percent of them are weak and 36.5 percent are strong. Table 4 also shows that just over half of the relations between non-firms are weak (55.3 percent) and just under half are strong (44.7 percent). Differences regarding the percentage of strong versus weak ties increase when we look at relations between firms. Just over three-quarters of the relations between firms are weak (78.5 percent) and just over one-fifth are strong (21.5 percent).

Table 5 shows the percentage of strong and weak ties between firms and non-firms. Firms have the highest proportion of strong relationships with research institutes. Among all the relations between firms and research institutes, 53.7 percent are strong and 46.4 percent are weak. On the other hand, the links between firms and other types of non-firm organizations are mainly weak: 75.4 percent of the links between firms and training institutions, 71.8 percent of the links between firms and government organizations, 56.8 percent of the links between companies and financial institutions, and 57.9 percent of the links between companies and local development support organizations are weak.

TABLE 5

Percentage of Strong and Weak Ties for the Relations between Firms and Different Types of Non-firm Organizations

	Weak	*Strong*
Ties With	*In % of ties*	
Research institutes	46.3	53.7
Training institutions	75.4	24.6
Government organizations	71.8	28.2
Financial institutions	56.8	43.2
Local development support organizations	57.9	42.1

Overall, our results indicate that interorganizational linkages within the Quebec optics/photonics industry are primarily weak. However, the social network literature argues that weak ties are not necessarily negative, because they can be an important source of new information, principally because these links are with people whom we rarely see (Granovetter 1973).

We also measured the "degree centrality" and "betweenness centrality" of each organization. The degree centrality is the total number of direct ties that an organization has in a network. The betweenness centrality measures the number of times an actor occurs on the shortest path linking a pair of actors (Freeman 1977, 1979). In other words, the betweenness centrality is a measure that captures the extent to which an organization is in a position to act as an intermediary.

Table 6 shows the top ten organizations according to degree centrality and betweenness centrality of the 58 organizations included in our study. For

TABLE 6
Top Ten Centrality Scores

Degree Centrality		*Betweenness Centrality*	
Organization	*Score*	*Organization*	*Score*
Research institute	54	Research institute	110.41
Financial institution	48	Local development support organization	64.10
Local development support organization	47	Company	45.22
Company	46	Financial institution	42.23
Local development support organization	44	Local development support organization	37.57
Company	42	Local development support organization	34.94
Local development support organization	41	Local development support organization	26.12
Research institute	39	Company	24.26
Company	39	Government organization	24.15
Training institution	38	Local development support organization	22.96
Company	38	Government organization	22.03

Note: Total number of actors is 58 companies.

confidentiality reasons, we list only the type of organization, not its name. We first note that the top ten members are not the same for the two measures of centrality, even if the index of betweenness centrality tends "to be highly correlated with degree centrality" (Bonacich, Oliver and Snijders 1998, p.135). A research institute is the most central organization according to both the degree centrality and the betweenness centrality. That we are dealing with a science-based cluster may explain the central position of this research institute. Finally, it must be noted that four firms are included in the degree centrality's top ten, and only two firms are included in the betweenness centrality's top ten. The two firms that are part of the betweenness centrality's top ten are also part of the degree centrality's top ten. The results suggest that the centre of the industry network is largely composed of non-firm organizations which are well positioned to coordinate information exchanges in the industry.

DISCUSSION AND CONCLUSION

These findings suggest that the Quebec optics industry has many important assets. First, the central place occupied by large public research organizations in the provision of research inputs, highly qualified personnel, and knowledge-brokers suggests that the Quebec industry is highly dependent on a strong, local, science-based cluster. Many of these research organizations are large public research organizations. By comparison, most American science-based clusters are based on industry research organizations. This suggests that large public research organizations may be particularly important in Canada for the success of science-based industries and help promote the emergence of science-based clusters.

Second, the Quebec industry is made up of a smaller number of firms than other optics/photonics clusters. These firms are smaller and rely more heavily on national capital than firms in other optics/photonics clusters. The Quebec firms occupy specialized niches of highly customized products with small markets. Furthermore, the customized products of the Quebec firms are closer to research results than is the case in most other optics/photonics clusters. This may suggest that the entry strategy of Canadian science-based firms is to occupy small markets with research-intensive products. It also suggests that Canada may be witnessing the emergence of a science-based industry model in which large, public research organizations feed small science-based firms to produce customized products for small markets.

If the Quebec optics industry and supporting organizations do constitute a cluster, the centrality of research organizations means that it does not rely on a large base of sophisticated local clients as Porter argues is usually the case. The Quebec firms are also part of world networks through clients with whom they work closely to co-develop highly complex and customized products/solutions. Again, it may suggest that small countries such as Canada have to rely primarily on clients from other countries to co-develop and purchase their products.

Fourth, our findings suggest that there are large gaps between facts and perceptions with respect to networking and clustering. When asked if they perceived themselves as being part of an informal network or cluster, only 44 percent of the actors involved in optics/photonics in the region said "yes" without hesitation. On the other hand, our sociometric analysis shows that numerous links exist between all of the actors in the Quebec region and that a vibrant network facilitates knowledge flows. This discrepancy suggests that actors involved in small science-based, export-oriented industries may have a narrow view of their environment and be blind to the local contributions to innovation.

Our findings suggest that there might be a Canadian model of science-based clusters that differs from the Porter model. The major characteristics of the Canadian model are the existence of a small vibrant network in which large public organizations work closely with a number of small firms competing for the co-development of customized, science-based products for the global market. Clearly, more research is needed to further document this model and to better understand its growth potential.

NOTES

We would like to thank Christine Forget for her participation in the project as well as all the participants without whom this project would not have been achieved. Their time, insights and enthusiasm are sincerely appreciated. The Quebec Optics Industry study is part of the Innovation Systems Research Network (ISRN) cluster study initiative. We would like to thank the SSHRC as well as David A. Wolfe and Meric S. Gertler from the ISRN for initiating and managing efficiently this project.

[1]Québec – Optics City's Web site at <http://www.quebecopticscity.com/en/business>. Consulted 26 May 2003.

[2]See COPL's Web site at <http://www.copl.ulaval.ca/english/center/historical.html>. Consulted 12 May 2003.

[3]INO's Web site for review. At <http://www.ino.qc.ca/en/a_propos/01_pro_1.asp#Fondation>. Consulted 12 May 2003.

[4]It is important to note that these numbers have been approximated and can vary depending on the definition used to define each category.

REFERENCES

Amara, N. and R. Landry. 2003. "Indicateurs de mesure des clusters d'innovation: une illustration à partir des données de l'enquête sur l'innovation dans les entreprises de Laval, des Laurentides et de Lanaudière." Study prepared for Canada Economic Development.

Bonacich, P., A. Oliver and T.A.B. Snijders. 1998. "Controlling for Size in Centrality Scores," *Social Networks* 20:135-41.

Borgatti, S.P., M.G. Everett and L.C. Freeman. 2002. *UCINET 6 for Windows*. Harvard: Analytic Technologies.

Brusco, S. 1990. *The Idea of the Industrial District: Industrial Districts and Interfirm Co-operation in Italy.* Geneva: International Institute for Labour Studies.

Braczyk, H.J., P. Cooke and M. Heidenreich, with G. Krauss. 1998. *Regional Innovation Systems: The Role of Governance in a Globalized World.* London: UCL Press.

Catts, B.C.1999. "Arizona Optics: A Targeted Industry Summary Report." Report prepared for the Business Development Division Arizona Department of Commerce.

Cowan, R. and J. Jonard. 2003. "The Working of Scientific Communities," in *Science and Innovation: Rethinking the Rationales for Funding and Governance*, ed. A. Geuna, A. Salter and W.E. Steinmueller. Northampton, MA: Edward Elgar, pp. 309-34.

Darr, E.D. and T.R. Kurtzberg. 2000. "An Investigation of Partner Similarity Dimensions on Knowledge Transfer," *Organizational Behavior and Human Decision Processes* 82(1):28-44.

de la Mothe, J. and G. Paquet, eds. 1998. *Local and Regional Systems of Innovation.* Amsterdam: Kluwer Academic Publishers.

Dicken, P., M. Forsgren *et al.* 1994. *The Local Embeddedness of Transnational Corporations: Globalization, Institutions and Regional Development in Europe.* Oxford: Oxford University Press.

Empire State Development Division of Policy and Research. 2001. "The Optics and Imaging Industry Cluster in New York State." Report prepared for the Empire State Development Division of Policy and Research.

Enright, M. 2000. "Policy for Inter-firm Networking and Clustering: A Practitioner's Perspective." Paper presented at OECD "Conference for Ministers Responsible for SMEs." Unpublished paper.

Freeman, L. C. 1977. "Centrality in Social Networks: Conceptual Clarification," *Social Networks* 1:215-39.

—— 1979. "A Set of Measures of Centrality Based upon Betweenness," *Sociometry* 40:35-41.

Goodman, E., J. Bamford and P. Saynor, eds. 1989. *Small Firms and Industrial Districts in Italy.* London and New York: Routledge.

Granovetter, M. 1973. "The Strength of Weak Ties," *American Journal of Sociology* 78(6):1360-80.

Hakansson, H. 1989. *Corporate Technological Behaviour: Co-operation and Networks.* London: Routledge.

Hendry, C. and J. Brown. 1998. "Clustering and Performance in the UK Opto-Electronics Industry." Presentation to "Conference on Regional Advantage and Innovation," Conference Universidade Catolica Portugese, Porto, 23-24 October.

Hendry, C., J. Brown and R. Defillippi. 2000. "Regional Clustering of High Technology-Based Firms: Opto-Electronics in Three Countries," *Regional Studies* 34(2):129-44.

Hendry, C. *et al.* 2001. "Innovating in Opto-Electronics in a Global and Regional Context." Paper presented at the 31st European Small Business Seminar, Dublin.

Holbrook, J.A. and D.A. Wolfe, eds. 2000. *Innovation, Institution and Territory. Regional Innovation Systems in Canada.* Kingston and Montreal: School of Policy Studies, Queen's University and McGill-Queen's University Press.

Landry, R. and N. Amara. 1998. "The Chaudière-Appalaches System of Industrial Innovation," in *Local and Regional Systems of Innovation,* ed. de la Mothe and Paquet, pp. 257-76

—— 2001. "Effects of Sources of Information on Novelty of Innovation in Canadian Manufacturing Firms: Evidence from the 1999 Statistics Canada Innovation Survey." Study prepared for Industry Canada.

Landry, R. and R. Gauthier. 2002. "Veille stratégique et étalonnage des modes d'organisation et de soutien du secteur de l'optique/photonique: un examen des pratiques régionales." Report prepared for GATIQ Technorégion.

—— 2003. "Rapport de veille sur les clusters: instruments et outils d'intervention." Working Paper prepared for Canada Economic Development. Unpublished paper.

Lawson, C. 1997. "Territorial Clustering and High-Technology Innovation: From Industrial Districts to Innovative Milieux." Working Paper No. 54. Cambridge: ESCR Center for Business Research, University of Cambridge.

Le Bas *et al.* 1998. "Innovation Technologique Comportement de Réseaux et Performances: Une Analyse Sur Données Individuelles," *Revue économique politique* 626-43.

Liyanage, S. 1995. "Breeding Innovation Clusters through Collaborative Research Networks," *Technovation* 15(9):553-67.

Lundvall, B.-Å. 1988. "Innovation as an Interactive Process-Form User-Producer Interaction to the National System of Innovation," in *Technical Change and Economic Theory*, ed. G. Dosi *et al.* London: Pinter Publishers.

—— 1992. *National Systems of Innovation: Towards a Theory of Innovation and Interactive Learning.* London: Pinter.

Maillat, D. 1995. *Systèmes territoriaux de production, milieux innovateurs et politiques régionales.* Université de Neuchâtel, Suisse: Version provisoire non publiée.

Mallett, J.G. 2002. "Silicon Valley North: The Formation of the Ottawa Innovation Cluster." Report prepared for the Information Technology Association of Canada.

Malmberg, A. and D. Power. 2003. "How Do Firms in Clusters Create Knowledge?" Paper presented at the DRUID Conference, "Creating, Sharing and Transferring Knowledge: The Role of Geography, Institutions and Organizations," Copenhagen, 12-14 June.

Marshall, A. 1920. *Principles of Economics*, 8th ed. Philadelphia: Porcupine Press.

Miyazaki, K. 1995. *Building Competences in the Firm: Lessons from Japanese and European Optoelectronics*. London: MacMillan.

Nauwelaers, C. and R. Wintjes. 2001. "SME Policy and the Regional Dimension of Innovation: Towards a New Paradigm for Innovation Policy?" MERIT, University of Maastricht.

Organisation for Economic Co-operation and Development (OECD). 1997. "Oslo Manual: Proposed Guidelines for Collecting and Interpreting Technological Data." Paris: OECD: Available at <http://www.oecd.org/LongAbstract/0,2546,en_2649_33703_2367514_119669_1_1_1,00.html>.

Porter, M.E. 1998*a*. "Clusters and Competition: New Agendas for Companies, Governments, and Institutions," in *On Competition*. Cambridge, MA: Harvard Business School Press.

—— 1998*b*. "Clusters and the New Economics of Competition," *Harvard Business Review* 76(6):77-90.

—— 2000. "Location, Competition, and Economic Development: Local Clusters in a Global Economy," *Economic Development Quarterly* 14(1):15-34.

Porter, M.E. and S. Stern. 2001. "Innovation: Location Matters," *MIT Sloan Management Review* 42(4):28-36.

Power, D. and M. Lundmark. 2003. "Working through Knowledge Pools: Labour Market Dynamics, the Transference of Knowledge and Ideas, and Industrial Clusters." Paper presented at DRUID Conference, "Creating, Sharing and Transferring Knowledge: The Role of Geography, Institutions and Organizations," Copenhagen, 12-14 June.

Putnam, R.D.1993. *Making Democracy Work*. Princeton, NJ: Princeton University Press.

Romijn, H. and M. Albaladejo. 2002. "Determinants of Innovation Capability in Small Electronics and Software Firms in Southeast England," *Research Policy* 31:1053-67.

Rosenfeld, S.A., 2002*a*. "Creating Smart Systems: A Guide to Cluster Strategies in Less Favoured Regions," *European Union-Regional Innovation Strategies*. At <http://europa.eu.int/comm/regional_policy/innovation/pdf/guide_rosenfeld_final.pdf>.

—— 2002*b*. *Just Clusters Economic Development Strategies that Reach More People and Places: A Synthesis of Experiences*. Carrbaro, NC: Regional Technology Strategies Inc.

Saxenian, A.L. 1994. *Regional Advantage: Culture and Competition in Silicon Valley and Route 128*. Cambridge, MA: Harvard University Press.

Scott, J. 2000. *Social Network Analysis: A Handbook*, 2nd ed. London: Sage.

Simmie, J. 1997. "The Origins and Characteristics of Innovation in Highly Innovative Areas: The Case of Hertfordshire," in *Innovation, Networks and Learning Regions*, ed. J. Simmie. London: Jessica Kingsley Publishers.

Sternberg, R. 1999. "Innovation Linkages and Proximity: Empirical Results from Recent Surveys of Small and Medium Sized Firms in German Regions," *Regional Studies* 33(6):539-40.

Storper, M. 1997. *The Regional World: Territorial Development in a Global Economy*. New York: The Guilford Press.

Suarez-Villa, L. and W. Walrod. 1997. "Operational Strategy, R&D and Intra-metropolitan Clustering in a Polycentric Structure: The Advanced Electronics Industries of the Los Angeles Basin," *Urban Studies* 34(9):1343-80.

Turner, R.C. 2001. *A Framework for Cluster-Based Economic Development Policies*. Albany, NY: The Nelson A. Rockefeller Institute for Government.

USF Office of Economic Development. 1999. "Report on Florida Laser and Optic's Cluster." Report prepared for the *Florida High Tech Corridor Council*.

Von Hippel, E. 1988. *The Sources of Innovation*. New York: Oxford University Press.

7

Clustered Beginnings: Anatomy of Multimedia in Toronto

John N.H. Britton and Gerry Legare

Background

The multimedia industry emerged in the mid-1990s as one part of the broad shift toward a digital economy and an increase in the importance of culture- and knowledge-based economic activity. It is a dynamic industry that encompasses a continually evolving mix of activities and products: producer/business services, information and communications technology, and the creative/cultural industries. The industry is unique in its aesthetic use of advanced graphics software platforms and in the generation of new digital applications that increasingly embody interactive content in terms of images, sound, and information. Although multimedia is not research-based, product design and development are crucial for the industry. The industry is highly creative and involves many forms of incremental innovation, including conceiving new applications of existing software and their implementation as new or improved products (see Scott 2000). The multimedia industry's products include complete Internet-based information systems, intermediate inputs for other visual media products, and entertainment and educational applications destined for the consumer market. In most cases, these new digital products meet what were futuristic expectations of graphics output only a few years ago. The industry serves an uncommonly wide range of clients from financial institutions through the production sector to entertainment producers and distributors, to government departments and educational institutions. Our interest in these characteristics has motivated this research.

Since it is extremely young as an industry internationally, the role of large-scale firms such as multinational companies has been experimental at best, and small firms remain one of the industry's dominant characteristics (PWC 2000; Spadina Bus Association 2001). These firms thrive partially because the technologies involved in many segments of the industry do not require large capital investments that would overwhelm their financial resources and management capabilities.[1] The speed with which multimedia concentrations emerged in major cities reflects a match between the rapid diffusion of software, hardware (particularly powerful desktop computers), and infrastructure and the availability of skilled and creative, young computer-graphics designers, communication experts, and programmers interested in developing interactive business and consumer applications. As with other producer service industries, multimedia emerged in key metropolitan centres because these locations provided ready access to a large pool of progressive users. Concentrations developed as computer-skilled entrepreneurs, technical workers and creative personnel moved into the industry from other fields such as software programming, publishing, and film and television production.

Despite its short history, the industry entered a fundamental restructuring phase after early 2001 following the collapse in the stock prices of digital and other high-tech industries. Although it is important not to equate multimedia firms with dot-com firms the two were closely linked. Dot-com firms were, and are, the clients of some multimedia firms (see Gorman 2002), but they focus on selling standard products and services, for example, books and travel on the Internet. The initial impact of a dot-com collapse on multimedia firms was to severely reduce their client pool. Reduced stock prices especially hurt firms that had accepted stock options in lieu of payment for services. The industry was further hurt because its close association with dot-coms damaged its business credibility. The decline in the business market for multimedia products was reinforced when the high technology sector went into recession. Conservative reactions to the continued downturn in North American markets after September 2001 (9/11), especially in the advertising sector, also reduced demand for multimedia services and products.

A third relevant factor is the increasing proportion of firms in all sectors that have new media departments or divisions. This trend is particularly evident among major advertising agencies, publishers, and entertainment producers. As a result, multimedia is beginning to reflect the organization of other producer services. Interviews confirm that product and service demands have dropped as clients stretch the cycle-length of projects and voice the need to ensure that investment in new forms of advertising and marketing will produce an acceptable return on investment.

CONCEPTUAL FRAMEWORK

Given Toronto's high concentration of entertainment production and its role as Canada's corporate centre, there is a strong case that its concentration of multimedia manifests considerable locational path dependence (Brail and Gertler 1999; Mills and Brail 2000). The question to be answered here is whether the geographically centralized locational pattern of many multimedia firms, which gave a spatially distinct identity to the industry about a decade ago, remains evident now. Alternatively, Toronto's concentration may be no more than an interesting co-location of firms among related economic activities (Wolfe 2003). One of our initial tasks, therefore, is to examine the degree of interdependency between Toronto's multimedia firms.

The focus of this chapter is on four forms of interdependence that commonly exist within industrial clusters. These involve the links between:

1. firms and the labour market;
2. firms and local or regional educational institutions;
3. firms involved in the production process; and
4. firms and business and civic associations.

These relationships address the issue of whether the cluster concept should be applied (Wolfe and Gertler 2003) to Toronto's multimedia concentration, which is still an emerging economic activity and for that reason is fragile and manifests uneven strengths. Nevertheless, it does exhibit many cluster characteristics. We examine these interdependencies through interviews with multimedia firms and a number of key informants familiar with the Toronto industry. The research agenda is approached as a case study problem, a methodology imposed by the lack of an official record of the multimedia industry. Therefore, we first consider the definition of multimedia and outline product divisions using sources primarily from within the industry. Interviews with firms are the primary source for the discussion of interdependencies.

Defining the Multimedia Industry

Multimedia products incorporate audio, video, graphics, and text elements delivered in digital form and the industry is the innovative outcome of creative content developers taking advantage of computer technology. For many multimedia firms and their clients, the Internet ultimately is the main distribution channel for their products (see also Eng and Patchell 2000), and therefore they rely on the infrastructure of the Web. This infrastructure is provided by a range of firms — Internet Service Providers (ISPs), portals, Web-hosting services, and Telecom and equipment suppliers — all of which lie outside our definition of the industry. There are alternative channels, however, for distributing multimedia products, such as intranets, kiosks, video terminals, CD-ROMs, and DVDs.

Multimedia firms producing for any of these channels use the same digital technologies, production processes, and labour skills. For this reason, with Dion (2001), we include visual animation and entertainment within the definition of multimedia; thus also embracing the production of interactive games, DVD films, video games, and on-demand films as part of the multimedia industry. On this basis, we also include firms whose products involve a wide array of 2D and 3D animation and digital special effects that go directly into film and television products (see also Lash and Wittel 2002).[2] These are also increasingly delivered via DVDs, Video-on-Demand (VOD), and similar systems, which have increasing potential to incorporate full interactivity.

We have adopted the following definition of multimedia:

> The digital media industry is actually the place where the computing, telecommunications and creative/content (i.e., the words, music, sound, visual images, stories) industries meet and create something fundamentally new and different. [It] relies on both the creative knowledge and skills of the content industry and the technical knowledge and skills of the computer industry to create a wide range of different products and services (Digital Media Champions Group 1998, p. 3).

We would add to this the characteristic that multimedia products are often used interactively or are embedded in interactive systems since this often enters other definitions (Lash and Wittel 2002).[3]

It is important to recognize that our definition focuses on combinations of creativity and innovation along a continuum from "pure" software and computational innovations to products whose novelty rests on creative content alone, regardless of the form they take. Content might be manifest in the form of interactive audio and video entertainment or in the form of a communications solution delivered for a business client through a Web site. Though the technological component of multimedia products is acquired predominantly through licensed software, much of the innovation in the multimedia industry involves the customization and implementation of systems design work and thus combines strategic and creative elements with technical expertise. Thus, we recognize a continuum of product creativity in terms of ideas for new products/services and innovation in terms of software tools built from code or within existing commercial platforms. In rare cases, early-stage firms develop new software systems before securing end-users.

Activities Included

Our general approach parallels Scott's (2000) work on California, where he describes two differentiated metropolitan multimedia clusters of firms: the Bay Area, which specializes in business applications and Los Angeles, which concentrates on entertainment content. As indicated above, there are grounds to expect that Toronto will resemble a small-scale version of the California industry. For this reason, we adopted a composite description that combines existing international literature on the enormous range of multimedia activities and core activities indicated by Toronto firms on their Web sites (Figure 1).

We expected the multimedia industry to have two product segments: business applications and entertainment applications. We verified this division through discussions with key informants from public agencies and local multimedia associations. We then modified the product subgroups in an iterative manner. As shown in Figure 1, the two spheres of multimedia activity are closely related to traditional business services and entertainment products. In some instances, the digital products we have included are substitutes for the conventional products but, most often, multimedia products offer new qualities and applications not previously available.

The dynamism of multimedia may be captured by conceptualizing the industry as two intersecting spheres of activity. Figure 1 incorporates our conclusion that regardless of the distinctions between the two market segments, they are continuing to converge and become increasingly interdependent in many segments, especially digital animation and visual effects. Animation, video, and interactive content are being increasingly embedded in e-business,

FIGURE 1
Product Structure of Toronto's Multimedia Industry

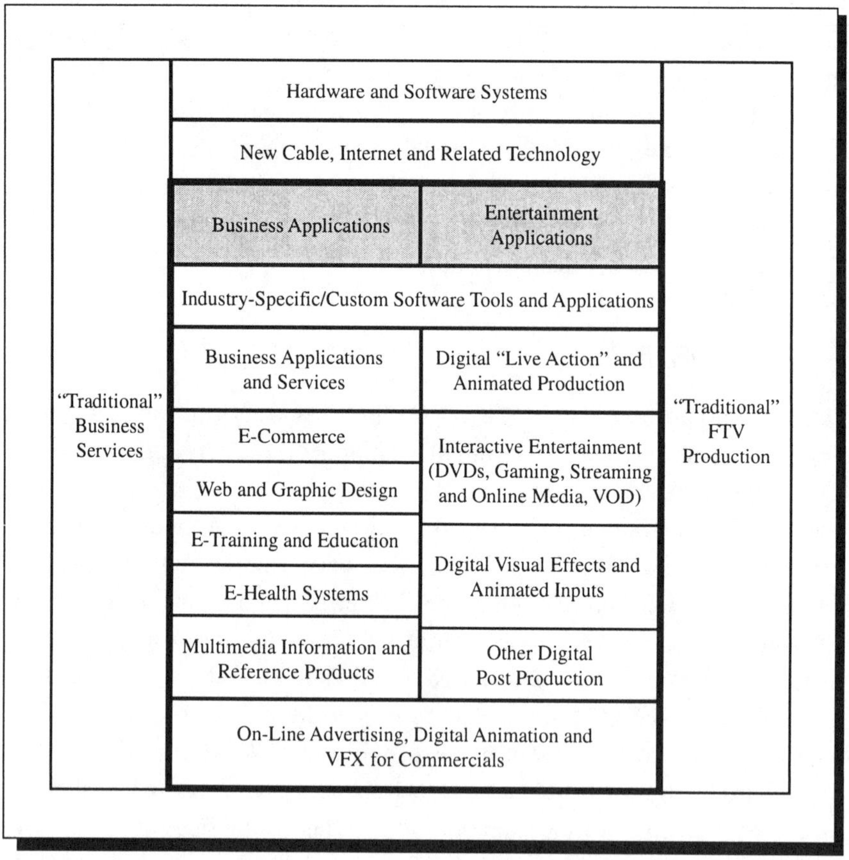

training or educational Web sites and the "live action" characteristic of traditional film and television (FTV) production is being increasingly incorporated into video games. This has led to the emergence of crossover firms that operate in both multimedia and traditional media sectors. Digital animation developed or applied in the production of entertainment content is also being used in online and interactive television advertising. Figure 1 categorizes this flow of entertainment content into business applications in an advertising product category that spans the two product segments.

Convergence is also reflected in a number of emerging markets, such as interactive television and online news and television feeds. Multimedia is a highly dynamic and rapidly evolving industry that defies easy differentiation and its new middle ground is some distance from its roots in film, television and audio entertainment, and telephone, fax and off-line database management. Figure 1 distinguishes six subgroups of business applications, most of which are online products, and four subgroups of entertainment applications most of which are commissioned as intermediate inputs for larger products. Despite this general pattern of demand, a variety of entertainment and educational products is available both on the Internet and in the form of DVDs or CDs.

Firms in both product segments use common infrastructure, similar technology and tend to license internationally marketed software platforms from local vendors under site-licence agreements. Some firms produce trademarked ™ software products in-house which are then licensed-out as part of interactive digital-service agreements. Since these firms bridge the divide between software production and content generation, Figure 1 shows the development of software tools as an activity of multimedia firms that spans the business/entertainment division. Infrastructure takes the form of systems of cable and Internet services and these also are important for both multimedia production and the development of distribution channels.

INDUSTRY PARAMETERS

Because of the predominance of project-based contracts in the multimedia industry it is difficult for firms to predict demand and thus to schedule workloads and allocate resources (see Grabher 2002). Projects may arrive sequentially or they may overlap in time but with different completion dates and dissimilar dimensions. This problem is multiplied by the variety of skill-sets and the variability of resource levels required for the production of custom products and services to meet each client's specific needs. These issues have strongly influenced the relationships that multimedia firms establish with their workforce, other firms, and clients.

The fee-for-service income model, under which most firms operate, also creates a number of management challenges that impact these relationships. The first challenge for firms is to establish fee structures that will generate sufficient revenues (Pollara Inc and Omnia Communications Inc 2000; Pratt

2000). This was less difficult during the economic boom of the mid-1990s, during which some multimedia firms in Toronto grew rapidly to more than 200 employees. Often, growth masked variations in company income and workforces were allowed to expand in anticipation of new projects.

Small firms, including branch establishments of the few multi-locational companies, are the norm in multimedia and market contraction has affected large firms as much as or more than small ones. Locating in multiple market centres through branch establishments failed to generate economies of scale in production may have created branding advantages for these firms. In their analysis of Silicon Alley (NY), Heydebrand and Mirón (2002) use ten employees as the dividing line between small and large firms. This also holds for Toronto where survey data (for 2000) show that 35 percent of multimedia firms employ less than ten full-time employees (FTE), with the mean size of this subgroup being less than four.

The recent struggles of many high-tech firms and generally tight markets for new business services has meant no growth (at best) for many of Toronto's multimedia firms. While demand has been sustained by conventional large firms and public sector clients for Web sites and other informational, advertising, and e-commerce services, smaller firms in all sectors have proved to be a less adventurous source of demand (see Statistics Canada 2003). Moreover, interviews indicate that multimedia clients have been stretching the time between projects and demanding better evidence that new forms of advertising, market development, and transaction management will produce an acceptable return on investment.

In contrast, shifts in the markets for film and television products have promoted both stability and adjustment in the entertainment segment. Over the last five years, in Canada and abroad, there has been a major increase in the number of television channels and subsequently in the demand for digital and animated program content. The positive impact on our subset of entertainment firms is the increase in the number of projects and their digital and/or animated components. Licence fees have fallen, however, putting pressure on firms to cut costs. This has reduced margins, but has made Toronto more attractive to US firms as a source of these products and inputs. Despite these shifts, there has been a palpable increase in the importance and sophistication of animated products for both television and film and of digital visual effects, primarily in feature film production and television advertising.

Scale of Multimedia in Toronto

There is little publicly available data on the multimedia industry in Canada, but industry reports estimate national employment at 20,000 FTE in 2002 (Delvinia Interactive Inc. 2003). This has dropped from more than 30,000 in 2001 (ibid.), in the face of an accelerated turnover of firms and employment. While most reports claim that approximately one-third of Canada's multimedia firms are located in Toronto, it is difficult to assess the size of Toronto's multimedia industry because there is no recognized database of firms. A 1997 study listed 319 firms in the Greater Toronto Area (Brail and Gertler 1999) while, currently, Industry Canada, the federal department responsible for industry, estimates there are 550 firms (Industry Canada 2004). Online and other membership lists and directories often contain out-of-date records, but by eliminating firms without current Web sites and adding firms that have emerged from interviews and other sources, our current list includes 685 firms. It is possible that the number of firms was greater in 2000 since we have encountered references to many that are now out of business.[4] It is also possible that the number of firms has grown because downsizing has increased the number of freelancers who list themselves as lone-proprietor firms.

A CLUSTERED LOCATION PATTERN

The international geographic literature on the multimedia industry indicates that firms are highly concentrated in a number of metropolitan locations (Cooke 2002). In Toronto, too, the fringe of the downtown business core emerged as a dense location of multimedia firms. Though the number of firms has increased and the economic activities in this area have diversified, it remains the largest concentration of firms. Three explanatory factors are important:

- The downtown concentration in Toronto of Canada's corporate offices, especially financial institutions, provides multimedia firms in the downtown fringe with excellent face-to-face access to very important clients.
- Toronto has a strong international identity as a film and television production centre. Many offices of film/television corporations are in downtown fringe locations making this district attractive for firms in the interactive entertainment segment of multimedia. Many of the founders of firms previously worked in conventional film, television and audio companies and are familiar with the area.

- There is a general concentration of cultural products industries in this geographic location, including art galleries, live music and recording, live theatre, fashion, and the arts in general, which influences both talent and the flow of ideas/inspiration.

These factors impact the multimedia industry in a variety of ways. Our interviews with industry personnel suggest that clustering in downtown locations has helped to establish the industry's identity by providing physical visibility. This applies to firms providing services for both the corporate market and the film and television industry. Their collective physical presence in Toronto contrasts with the ephemeral impression left by dot-coms. Low rents and the attraction of renovated nineteenth-century brick-and-beam buildings were important in the early 1990s. Later in 1996, zoning changes increased the transformation of industrial warehouses into residential "lofts," entertainment and other design-intensive activities, thus driving up rents. A small number of firms have located in more conventional office buildings, though still within the central downtown area. Their philosophy has been to reflect the office design norms of their business clients.

From the standpoint of attracting a young, talented workforce, good transit connections and the nearby location of clubs and restaurants are particular advantages. Moreover, the ease of interaction between workers in different firms increases the sharing of information about job opportunities. The small scale of firms in Toronto's multimedia industry and their agglomeration in the central area are reminiscent of vertically disintegrated industrial districts, which have been recognized as a modern component of metropolitan areas just as they were in earlier industrial cities (Hayter 1997; Scott 2000; Rantisi 2002). It is still an open question though, whether the industrial district model really fits the multimedia industry.

THE WORKFORCE AND THE PROJECT NATURE OF WORK

Multimedia's young, design-conscious and "trendy" labour force exerts a strong influence on the location of many firms. In this respect the "cool" factor is associated with a central location, age-related entertainment (bars, clubs, music), and the reuse of brick-and-beam style of former industrial buildings. This geography contrasts with the largely suburban location of college programs that feed the labour market for multimedia firms. The dot-com meltdown, which the founders and managers of most multimedia firms closely experienced, has also shaped the industry. Human-resource priorities include concern for the family life-cycle interests of workers who helped establish the industry. Firms attempt to retain loyal and skilled employees by trying to stabilize employment through balancing the workflow of projects in accordance with production capability. In other words, they aim for workforce stability in the context of inherent instability. This goal is partly socially constructed and partly a reflection of business self-interest in maintaining a skilled workforce. Some firms view this as a less important issue and have not kept the year-over-year flow of work in line with employment. A common practice is to rely on a contingent labour force hired on project-related contracts. While some of these workers are former full-time employees and others are new entrants to the industry, most are seeking full-time positions.

The boom-and-bust cycle caused many workers to leave the industry. Others have positioned themselves as freelance contractors, often working from home offices, though often they are physically located within firms as members of project teams. Christopherson (2002) notes that freelance workers may think of themselves as entrepreneurs or "free agents." This is justified because the most valuable workers, in fact, are often highly skilled freelancers who can achieve higher incomes by subcontracting projects from a variety of firms. They are "consultants" in a wide range of disciplines from software specialties to market strategy. Nevertheless, looking for new work often incurs very high transaction costs and effective personal networks are needed to reduce these costs.

While one can distinguish workers according to employment status, it is difficult to differentiate them according to market segment. Because the same software platforms are used in closely related areas, such as cartoon animation and web animation, there is little difference between the skills required. It is therefore possible to view multimedia firms operating in one labour market. Before 2001 firms grew by hiring workers away from others. However, the industry has generated a rich pool of talent through on-the-job learning and there is significant labour mobility within the region because of the nature of the secondary labour market. In recent years, the labour pool has been enriched by a small but steady flow of highly experienced, Canadian talent returning from the US ready to raise families and work "at home."

Despite the depth of the labour market, interviews indicate that firms strongly prefer to hire workers with experience and a proven ability to produce. This is particularly relevant when senior positions have to be filled. The present difficulties some firms are having in filling these positions may be due to the preference of many experienced workers for freelancing. Just as freelancers require personal networks to find work, firms need them to acquire the specific creative capacity and technical skills required for particular contracts. This pattern of behaviour requires effective personal networks (see Christopherson 2002). These same networks are of primary importance in the way firms acquire creative capacity and technical knowledge inputs via skilled specialist workers. The links in the networks that we explored in interviews have an almost exclusive local geography. Though a few firms use "consultants" from outside the region, without exception they indicated their overwhelming reliance on their personal file of specialists, and these networks are almost exclusively local.

UNIVERSITY AND COLLEGE LINKAGES

The increasing numbers of students graduating from new university and college programs in multimedia applications are creating an excess of inexperienced workers at a time when wages are low. Though these graduates are coming from specialized programs in animation, computer and software engineering, and other digital specializations, many of our interviewees argued that the most talented workers are self-taught. Despite skepticism about the value of specialized training programs versus learning-by-doing, industry personnel believe that colleges provide an important point where creative talent and technical disciplines can be systematically combined.

The majority of firms acknowledge that these programs produce entry-level talent with basic skills and help to sustain the local labour force. Nevertheless, few relationships exist between firms and educational and training institutions, and firms indicated that building these relationships was not on their agenda. The vast majority lack ties to local programs and have a variety of reasons for this. Primarily, it is a matter of not having the personnel resources to establish these relationships — especially for smaller firms facing tight budgets, close deadlines, and irregular work schedules. Most find it difficult to release key workers to participate in program planning or advisory boards when in the past this has not produced tangible returns. Moreover, there is a pervasive opinion that the educational system is unresponsive to the needs of firms.

Six of the firms we interviewed, however, have active relationships with local and regional university and college programs.[5] This usually involved a member of the firm acting as an instructor in a course, giving guest lectures, or sitting on a program committee. These are seen as both a way of "giving back" to the community and an opportunity to scout for talent. We anticipated that firms would host co-op students, but financial constraints have prevented this, because allocating workspace, software licences and rendering time can require investments of anywhere between $50,000 to $75,000.

INTER-FIRM INTERDEPENDENCE

The importance of local personal networks in the labour market operation suggests that Toronto's multimedia firms likely have a parallel set of local inter-firm relations in production (Christopherson 2002). These connections have been forged through both repeated working arrangements and familiarity between account/project managers and other executives. A firm's willingness and flexibility in partnering on new projects often deepens these relationships and provides a means of disseminating knowledge through the industry. These are common characteristics of clusters (Asheim 1996; Ekinsmyth 2002).

With its emphasis on highly localized interactions, this version of the cluster model is a poor fit for inter-firm networks that rely heavily on knowledge inputs and niche markets which are international in scope (Markusen 1996; Oinas 1999). A number of studies of concentrations of manufacturing firms in southern and central Ontario, for example, demonstrate the openness of clusters to inflows of capital equipment, components, software, other forms of technology, and knowledge from outside the region. International markets are also often more important than domestic markets for Canadian firms in many high-tech industries (MacPherson 1987; Anderson 1995; Cornish 1997; Gertler, Wolfe and Garkut 2000; Britton 2003). For this reason, the degree to which multimedia relies on domestic markets and local inter-firm relationships is one of the questions this study attempts to answer.

The small scale of individual firms (especially on the business side) means that there are no anchor or hub firms that might help to establish a national or international identity for the industry. This absence means that there are no subcontracting arrangements managed by a hub firm. Moreover, an earlier consultant's report (PWC 2000) concluded that there are low levels of connection between firms. Nevertheless, we have explored the possibility of inter-firm relationships, especially those formed as firms seek to combine their respective specializations. We view these as components of production networks, recognizing that the literature on industrial clusters focuses on links between firms that result from supply chains, business associations, and training and research institutions.

Vertical Interdependence

Information on the functional connections established by multimedia firms is drawn solely from 25 in-depth interviews, mainly with firms of above median size.[6] A composite strategy was developed in selecting firms to interview. References to innovative firms were collected from previous research papers and the business press and from discussions with innovation advisors and representatives of multimedia associations. The names of additional companies emerged in the course of interviews, as competitors and members of consortia were identified. The companies selected in this way tend to be older (nearly two-thirds of the business market firms and all of the entertainment firms are seven or more years old). These firms are also larger than the industry norm (mean employment size of 30 for the business market firms and 66 for the entertainment-focused group). Because all firms had survived the dot-com bust and the rigors of downsizing, they could offer thoughtful reflections on that experience.

Business applications segment. The forward connections of multimedia firms have strong locational relationships with the head offices of their corporate clients (Scott 2000; Lash and Wittel 2002). Web-site listings of such clients show that this is also the case in Toronto. Interviews indicate a number of benefits associated with such proximity. Face-to-face meetings are considered important to negotiate contracts, assess client satisfaction, and to develop a continuing client relationship. With local sales accounting for 64 percent of multimedia business, these links are clearly important. Only two very small firms generate international sales above 50 percent. While many feel that the exchange rate gives Canadian firms an important cost advantage when bidding on US contracts, they believe that other factors are more important. Securing international clients helps firms enhance their local reputation by showing that they can meet the technical and creative challenges of accessing the international market. There is some evidence that demanding local clients (fast adopters) enabled Toronto firms to develop applications that provide them with the tools to enter the US market.

There is no standard procedure by which firms secure clients. Almost all firms take referrals from advertising agencies, public relations firms, and consultants of various types and respond to Requests for Proposals (RFPs). This confirms that much of the industry's output is project based — or at least begins on this basis. Projects are sometimes completed by single firms (usually the most experienced firms), but many out-source work to other firms. Subcontracting is often a matter of skill acquisition and convenience, but it can be less costly than acquiring and managing a temporary team, allocating space, and supervising the work. About half the firms interviewed have been drawn into projects by other, usually larger, firms needing specialists able to provide the functional "back-end" technology that makes a Web site interactive. There is, however, a joint bidding process that allows firms with equal experience, even if they differ in scale, to work as a consortium.

Project revenue generates an episodic flow of income to firms but exists alongside more extended relationships with clients. Some of these resemble the agency of record relationship used in the advertising industry, which is assigned to an agency to manage ongoing media access and content so that a longer-term service relationship may be developed. Continuing relationships of this type are highly prized because they provide a stable revenue stream. All the firms studied have some relationships of this type; often involving an "understanding" that related new work will go to the firm without competitive bidding. These arrangements, however, do not preclude interventions by other firms seeking to bid on future work.

In addition to their production-based relationships, business-market multimedia firms rely heavily on a wide variety of inputs from local specialist firms: Web-hosting services, usability studies, video, animation and post-production inputs, educational content, and game components for Web sites. Local agents for proprietary software platforms are also important. Some of the more specialized proprietary software, which may be used as modules in larger applications, are acquired locally, but some tasks, such as security and encryption involve direct imports. Many firms purchase market reports from US media-watch agencies such as Forrester Research and Jupiter Research.

Multi-firm contract bids are the best evidence we have that firms strategically combine specialist strengths on a regular, if not ongoing basis. Underlying these collaborative initiatives are trust- and performance-based relationships between the individuals designing the bids. These arrangements recognize the strengths of specialist firms, but also the impossibility of their making a plausible lone bid for certain projects because of their scale. In many other industries firms may merge to avoid competitive limitations of small scale. In this industry this option is not likely since most firms are private and seek stability rather than rapid growth. It is also questionable whether multimedia firms have intellectual property assets that can be acquired through merger. Nevertheless, there are some cases in Toronto where multimedia firms have been acquired so that other firms (not always local ones) could obtain their technology, client lists, or personnel skills. There are other instances in which advertising agencies invested in Web-design firms so as to access their capacity/expertise on a preferential basis while others have absorbed multimedia firms to avoid the costs of developing an in-house unit from scratch.

Entertainment applications segment. The entertainment side of Toronto's multimedia industry includes the digital component of the film and television industry, gaming, and other Web-based forms of entertainment. Many of the technical innovations in the film and television industry in the last ten years have been in the areas of animation, visual effects, and post-production. Digital technologies continue to revolutionize the way film and television programs and television commercials are produced. Advances in computer hardware and software enable live-action and digital characters and backgrounds to interact seamlessly.

Toronto's substantial film and television production industry has directly influenced the emergence and growth of (multimedia) firms specialized in digital animation and visual effects. Interviewees claimed that their relationships with local clients and suppliers are more stable and enduring than non-local linkages. However, they also revealed that a considerable volume of work is undertaken for US clients. The favourable exchange rate is a particular advantage for Toronto firms. Because work for US companies, on average 45 percent of output, is much more sophisticated and technically challenging than work for the Canadian market, many firms consider it more important than domestic work despite their closer local relationships. This work involves much larger production budgets and is a stronger stimulus to the development of technological and creative skills and innovation. Firms regard linkages with creative and demanding clients as especially important sources of knowledge transfer and a spur to innovation. This is especially true of foreign collaborators and "experts," who often spend little time in Toronto, and whose impact may occur only while they are in Toronto.

Whether clients are local or international, interviewees indicated that personal relationships between key individuals have a very important role in securing contracts within the context of previous interactions between the firms. In domestic work, however, budgets are likely to play as critical a role. In part, this is due to labour mobility, but it also reflects the importance of individual creativity in this sector. Multimedia firms active in the entertainment and business markets exhibit a highly individualized variant of performance-based trust (Cohen and Fields 1999). Firms serving the entertainment market appear to be less active in forming consortia or entering joint bids for contracts than those serving the business market. Subcontracting does occur though, and firms in visual effects, animation, and visual post-production often have long-term formal or informal relationships with audio production and post-production companies. In this way, they are similar to firms in the business market segment. In subsequent research, it will be important to verify that this result is not scale related, but reflects only issues of reputation and specialization. Some firms in this segment are "tied" through ownership to film, television, and other production firms; in these cases, of course, there is ample opportunity for subcontracting.

In terms of their inputs, entertainment firms depend on some of the same software/ hardware technologies as the rest of the multimedia industry. However, they also use more advanced animation software, such as *Maya*, and invest in more powerful hardware systems. Their increasingly international market requires these firms to be at the leading edge of new hardware and software innovations. Software upgrades for most of the key systems function on a 10 to 12 month cycle, with intermediate updates occurring on a regular basis. The most competitive firms must be in nearly constant contact with their suppliers as they experiment with interface elements and press for "workarounds" for project-specific functions. This introduces considerable support costs.

A substantial amount of design work on new software tools is undertaken in-house and should be viewed as a form of software R&D. The same is true of their recent advances in workflow optimization, which has allowed many firms to cut both the physical and time costs of production. It is, therefore, impossible to overestimate the importance of labour skills in programming and in creative work among both full-time and contract workers. On-the-job experience seems essential, and freelance workers gain valuable knowledge from their exposure to a variety of workplaces and intense interactions with workers in animation and other specialties (Norcliffe and Eberts 1999). Their circulation between projects and firms then acts as an important mechanism of knowledge transfer.

Associational Relationships

Toronto's multimedia industry has generated a number of business and professional associations, each of which claims its own mission. However, many have complained about the lack of a single voice for the industry. This reflects similar trends in the film and television sector, where the lack of a coherent voice is continuously lamented. Membership lists and directories make it clear that relatively distinct sets of firms have tended to join different organizations and treat them as alternatives. The large number of organizations may reflect the industry's youth and instability, but it may also indicate that many needs are not being met.

Interviewees complain that associations are weak, unfocused and too numerous to command attention. For firms, there are too many events and too few occasions to connect with prospective clients. These responses may reflect the more established position of the firms that have been interviewed to date. Nevertheless, this assessment does not apply to associations created for workers, many of whom are freelancers or only intermittently employed. Lacking regular interaction with colleagues these individuals need opportunities to share and gather information about changes in the industry. Usually, however, associations attempt to meet the needs of both firms and individuals. Connection-forming and knowledge-pooling-and-sharing functions dominate.

There has been a recent trend toward the consolidation of multimedia-related organizations within the Toronto region. In particular, Smart Toronto, an association with a broad mandate, absorbed the Liberty Village New Media Centre in 2002 to form Smart Toronto Technology Alliance (STTA). This drew in other groups such as OnTarget (whose mission is to encourage high school students to enter media programs), the Toronto New Media Trainers Alliance (which links college level multimedia programs) and the Toronto chapter of DigitalEve. In 2003, STTA became a division of the Canadian Advanced Technology Alliance (CATA). This arrangement was intended to enhance the visibility and promote the policy needs of Toronto multimedia firms on a national scale. It remains to be seen if this will succeed. There is also the possibility that the multimedia industry's voice will be lost within the association's broader high technology mandate. Further, it is unclear how this development will strengthen the linkages between firms focused on the business and entertainment market segments.

An alternative trend has seen the emergence of associations catering to the particular interests of emerging fields like digital game production and online marketing. The Association of Internet Marketing and Sales (AIMS), for example, received overwhelming support from our interviewees. One of the association's goals is to advance knowledge of best-practice online marketing, which it does by showcasing peer-evaluated products for potential clients. Among industry organizations, AIMS provides the clearest indication that Toronto firms recognize the potential benefits of cooperative interaction with their competitors. This provides a formal structure for firms to learn by "watching, discussing and comparing dissimilar solutions" (Maskell 2001, p. 929). This approach to the circulation of knowledge does not hold, however, for animation and visual special effects firms, which are much more informal in their relationships and generally more individualistic.

Relationships with Government

Despite high levels of interest from both parties, relationships between the federal and provincial governments and the multimedia industry are still in the formative stage. Where programs exist, they have evolved out of existing programs for the technology or the cultural industries or are focused on education and training, and for the most part they continue to be poor fits for the needs of the industry. Firms focused on the business-market generally receive no direct public support except for research and development. When firms do undertake experimental product development they rely heavily on assistance from the federal and provincial governments, but we have found great variance in their knowledge about the kinds of assistance available for this activity. There is an obvious need to forge a closer relationship between potential innovators and the programs available to them.

Entertainment-focused firms have access to a variety of public and private grants for new cultural products. The federal government has two major programs that offset some of the costs in developing creative content. Telefilm Canada (Canada New Media Fund) and the Canada Council for the Arts administer direct support for multimedia, while the Bell (Broadcast and New

Media) Fund disburses revenue received from Bell ExpressVu under a requirement set by the federal regulatory body for radio, television and telecommunications (CRTC).

The Ontario Media Development Corporation is also a potential source of support for many multimedia firms with creative and identifiably Canadian ideas for new digital products. It is difficult for most firms to access these funds, however, because the small size of the Canadian market and the amount of funding available make it difficult to produce a commercially viable product while respecting the Canadian content requirement. As well, the program stipulates that producers retain intellectual property rights and not transfer them to distributors/publishers/co-producers. This conflicts with the usual industry practices in digital games and broadcasting. For all of the programs at the federal and provincial levels, the mismatch between application requirements, production timelines, and the availability of funds is also a major issue, given firms' general lack of capital resources. At this stage we are unsure if Toronto's multimedia industry, particularly the entertainment sector, can achieve a stronger position in international markets if the only assistance strategy in place is one that is focused on Canadian content in the manner adopted by Ontario and the federal governments. It is also clear that there is no coherent development policy that meets the needs of the entire multimedia industry at this point.

IS THIS A CLUSTER?

Given the spatial concentration of the multimedia industry in Toronto and its localization within the fringes of the central area, we began our research with the hypothesis that we would find support for this being a functional cluster. We have found good evidence to support this interpretation. The speed with which this infant industry is changing means there is a need for constant monitoring of new developments. In particular, locations outside the central area are proving suitable though they, too, have benefited from the scale of the industry's development in Toronto.

One source of the cluster's strength is the large number of local support firms and infrastructure investments linked to the industry, including software vendors, Web-hosting servers, and other infrastructure. In particular, the bandwidth of cabling has improved quickly and there have been localized spillover effects of large and financially successful cable users. Nevertheless, the cluster's chief strength derives from its internal bonds and the indicators we use — vertical interdependencies between firms, their activities in professional associations, and the existence of extensive personal and professional networks — show that firms and individuals increase their knowledge and their innovative capabilities through such interdependencies.

We began our inquiry with a clear sense of the product differences between two main segments of the multimedia industry. We found that there is also convergence between them. In particular, animation and visual special effects firms are increasingly interested in online applications and products. It is highly likely this trend will be reinforced as digital technology is increasingly incorporated into entertainment products. The two segments of multimedia are not geographically distinguishable in any way, and the use of local personal networks to acquire labour and to secure contracts with clients seems to span the product categories we have identified. Repeat connections are the norm, and firms not only make regular use of consultants/freelancers, but also attempt to balance workflow so as to retain core (long-term) employees. Many firms, which have downsized, now use former employees as freelancers on a project basis. This employment pattern has created new linkages between multimedia firms, facilitating the flow of knowledge.

Access to this highly experienced and qualified labour force is one of the most important factors determining the capabilities of individual firms and the multimedia industry as a whole because of the importance of individual talent, skill, and knowledge. There is an oversupply of new recruits, now graduating from postsecondary educational institutions, who tend to be inexperienced because internships and cooperative programs remain scarce and are rarely paid. Nevertheless, the New Media Trainers Alliance argues that strong, local educational programs are one of Toronto's important advantages. Ultimately this may be true, though at this time we find limited evidence of a pattern of connections/linkages between colleges or universities and firms. Rather, firms have found it difficult to become involved due to financial and resource limitations and as most firms do not undertake research-based projects they see little benefit from such connections. This lack of connection is understandable given that firms are mainly small and market conditions dictate that they devote all resources to defining and retaining their core competence.

Small scale probably also lies at the root of the significant disaffection among firms with industry associations and government programs. Programs appear to offer support to firms but are a source of frustration over their narrow application and their fragmentation. Meanwhile, associations, which ostensibly are the creations of firms in the industry, rarely deliver exactly what firms seem to want. These are issues that must be addressed for the ongoing development of the cluster.

The most telling area of identifiable cluster strength is probably the links of vertical interdependence between firms that are forged on an everyday basis. Though these are well established, there is clearly variation in their form and intensity between market sectors. Many of the firms serving the business market are similar with respect to their client relationships to other producer service firms, especially those in consulting, advertising, and marketing. In contrast, firms serving the entertainment market, such as those in animation and visual special effects, are an integral part of larger "collaborative" entertainment production systems in the film and television industry. They also produce stand-alone products for the consumer market. In both sectors, firms occasionally subcontract work to reduce internal workflow congestion. Firms serving the business market, however, more frequently collaborate on project bids and are more adept at pooling resources with other firms at this and later stages of projects. Some firms are known for their more aggressive, competitive style, but more commonly firms maintain a high level of intelligence about the activities of their competitors, through formal and informal means. They rely on personal connections to extend peer relationships and develop collaborative relationships. They use effective personal relationships as much as professional ones to secure subsequent contracts with clients. Given the importance of these characteristics, it is clear that the success of multimedia in Toronto may be understood by employing some of the subtler network elements of the cluster model.

NOTES

[1]In specific areas such as the leading edge of visual and special effects costs are much higher because of the need for memory and processing power in tasks such as rendering sophisticated graphics, and firms tend to be marginally larger.

[2]In some earlier Canadian literature, firms specializing in animation, post-production, and visual effects were excluded (Mills and Brail 2000) but this may be matter of timing.

[3]We have not adopted alternative industry names such as "interactive media" or "new media" (Pratt 2000). Names are a matter of fashion (Cooke 2002) and there is no standard.

[4]Multimediator (online), IMAT (online), Strategis (online), PWC (2000), Spadina Bus Association (2001), and City of Mississauga (2002).

[5]Two firms were in the business segment, three were in the entertainment segment, and one was mainly a "software" producer.

[6]There were 14 interviews in the business applications segment and 11 in the entertainment segment.

REFERENCES

Anderson, M. 1995. "The Role of Collaborative Integration in Industrial Organization: Observations from the Canadian Aerospace Industry," *Economic Geography* 71:55-78.

Asheim, B. 1996. "Industrial Districts as 'Learning Regions': A Condition for Prosperity," *European Planning Studies* 4:379-400.

Braczyk, H.-J., G. Fuchs and H.-G. Wolf, eds. 1999. *Multimedia and Regional Economic Restructuring*. London: Routledge.

Brail, S.G. and M.S. Gertler. 1999. "The Digital Regional Economy," in *Multimedia and Regional Economic Restructuring*, ed. H.-J. Braczyk, G. Fuchs and H.-G. Wolf. London: Routledge.

Britton, J.N.H. 2003. "Network Structure of an Industrial Cluster: Electronics in Toronto," *Environment and Planning A* 35:983-1006.

Christopherson, S. 2002. "Project Work in Context: Regulatory Change and the New Geography of Media," *Environment and Planning A* 34:2003-15.

City of Mississauga. 2003. *Mississauga Business Directory*. Mississauga: Mississauga Economic Development Office.

Cohen, S.S. and G. Fields. 1999. "Social Capital and Capital Gains in Silicon Valley," *California Management Review* 41:108-30.

Cooke, P. 2002. "New Media and New Economy Cluster Dynamics," in *Handbook of New Media*, ed. L.A. Lievrouw and S. Livingstone. London: Sage Publications.

Cornish, S. 1997. "Product Innovation and the Spatial Dynamics of Market Intelligence: Does Proximity to Markets Matter?" *Economic Geography* 73:143-65.

Delvinia Interactive Inc. 2003. *Interactive Media Producers Survey 2002*. Available at <www.delvinia.com/IMPS2002_English_Exec_Sum.pdf>.

Digital Media Champions Group. 1998. *Playing to Win: The Digital Media Industry in Ontario*. Available at <www.multimediator.com/dmcg/Playing_to_Win.pdf >.

Dion, S. 2001. *Profile of the Canadian Multimedia Industry*. Montreal: Interactive Multimedia Producers Association of Canada, for Industry Canada.

Ekinsmyth, C. 2002. "Project Organization, Embeddedness and Risk in Magazine Publishing," *Regional Studies* 36:229-43.

Eng, I. and J. Patchell. 2000. "University-Industry Research Centres and Regional Development: Matching Applied Research to Hong Kong's Nacent Multimedia Industry," *Regional Studies* 34:494-99.

Gertler, M., D. Wolfe and D. Garkut. 2000. "No Place Like Home? The Embeddedness of Innovation in a Regional Economy," *Review of International Political Economy* 7:688-718.

Gorman, S.P. 2002. "Where Are the Web Factories: The Urban Bias of E-Business Location," *Tijdschrift voor Economishe en Sociale Geografie* 93:522-36.

Grabher, G. 2002. "Cool Projects, Boring Institutions: Temporary Collaboration in Social Context," *Regional Studies* 36:205-14.

Hayter, R. 1997. *The Dynamics of Industrial Location*. Chichester: Wiley.

Heydebrand, W. and A. Mirón. 2002. "Constructing Innovativeness in New-Media Start-up Firms," *Environment and Planning A* 34:1951-84.

Industry Canada. 2004. Toronto New Media – Industry Profile. Available at <http://strategis.ic.gc.ca/epic/internet/inict_c-g_tic.nsf/en/h_tk00184e.html>.

Lash, S. and A. Wittel. 2002. "Shifting New Media: From Content to Consultancy, from Heterarchy to Hierarchy," *Environment and Planning A* 34:1985-2001.

MacPherson, A. 1987. "Industrial Innovation in the Small Business Sector: Empirical Evidence from Metropolitan Toronto," *Environment and Planning A* 20:953-71.

Markusen, A. 1996. "Sticky Places in Slippery Space: A Typology of Industrial Districts," *Economic Geography* 72:293-313.

Maskell, P. 2001 "Towards a Knowledge-based Theory of the Geographical Cluster," *Industrial and Corporate Change* 10:921-43.

Mills, L. and S.G. Brail. 2000. "New Media in the New Millennium: The Toronto Cluster in Transition," in *Knowledge, Clusters and Regional Innovation*, ed. A.J. Holbrook and D.A. Wolfe. Kingston and Montreal: School of Policy Studies, Queen's University and McGill-Queen's University Press.

Norcliffe, G. and D. Eberts. 1999. "The New Artisans and Metropolitan Space: The Computer Animation Industry in Toronto," in *Entre la métropolisation et le village global*, ed. J.-M. Fontan, J.-L. Klein and D.-G. Tremblay. Sainte-Foy: Presses de l'Université du Québec.

Oinas, P. 1999. "Activity-Specificity in Organizational Learning: Implications for Analyzing the Role of Proximity," *GeoJournal* 49:363-72.

Pollara Inc and Omnia Communications Inc. 2000. *Ontario Digital Media Production Activity Profile*. Toronto: Ministry of Citizenship, Culture and Recreation and Ontario Film Development Corporation.

Pratt, A.C. 2000. "News Media, the New Economy and New Spaces," *Geoforum* 31:425-36.

PricewaterhouseCoopers (PWC). 2000. *Toronto New Media Works Study*. Toronto: PricewaterhouseCoopers.

Rantisi, N. 2002. "The Local Innovation System as a Source of 'Variety': Openness and Adaptability in New York City's Garment District," *Regional Studies* 36:587-602.

Scott, A.J. 1998. "From Silicon Valley to Hollywood," in *Regional Innovation Systems*, ed. H.-J. Braczyk, P. Cooke and M. Hiedenreich. London: UCL Press.

—— 2000. *The Cultural Economy of Cities*. London: Sage.

Spadina Bus Association. 2001. *Route Map: A Snapshot of Toronto's New Economy*. Toronto.

Statistics Canada. 2003. "Electronic Commerce and Technology 2003." *The Daily*, 2 April.

Wolfe, D.A., ed. 2003. *Clusters Old and New: The Transition to a Knowledge Economy in Canada's Regions*. Kingston and Montreal: School of Policy Studies, Queen's University and McGill-Queen's University Press.

Wolfe, D.A. and M.S. Gertler. 2003. "Clusters Old and New: Lessons from the ISRN Study of Cluster Development," in *Clusters Old and New*, ed. Wolfe.

8

The Montreal Multimedia Sector: District, Cluster or Localized System of Production?

Diane-Gabrielle Tremblay, Catherine Chevrier and Serge Rousseau

Introduction

Along with Vancouver and Toronto, Montreal is home to one of the principal concentrations of Canada's multimedia or "new media" industry. Over the last decade this industry has grown significantly and has attracted much attention. This chapter presents the preliminary results from our study of Montreal's new media industry. We begin by briefly defining the concepts of "industrial districts" and "clusters," which underlie our research, and examining some definitions of the multimedia sector. We then describe Quebec's multimedia sector, focusing in particular on the characteristics of its various actors as well as its labour force, one of the key elements in most descriptions of clusters or industrial districts. We conclude by presenting a few preliminary observations on the role that interorganizational relations and government financial support have played in the industry's development and by discussing whether or not the Montreal multimedia sector should be defined as a cluster, a district, or a localized system of production. We base most of our analysis on documentary research and 75 interviews conducted with industry personnel in the ISRN project. In addition, some of our findings draw on approximately 50 interviews with multimedia personnel carried out during the Telelearning Centres of Excellence project (2000–2004).

DISTRICTS AND CLUSTERS

We begin by examining the concepts of "industrial districts" and clusters. These concepts are useful for our research because our overall goal is to study the role of inter-firm relations in innovation and territorial development. In recent years a number of theories, particularly those describing innovative milieux and localized systems of production, have addressed how geographical proximity between firms impacts innovation (Klein, Tremblay and Fontan 2003; Klein, Fontan and Tremblay 1999, 2001; Fontan, Klein and Tremblay 2001). We find that proximity has strongly influenced the development of Montreal's new media industry, arguing that the Cité du multimédia project in Montreal is a form of "organized proximity" that builds on the geographic proximity of firms in Montreal. In addition, we find that the provincial government has encouraged firms to move into Montreal by offering them financial incentives.

Industrial Districts

Alfred Marshall put forward the concept of an industrial district at the end of the nineteenth century. It surfaced again in work by Italian economists studying the so called "Third Italy," a number of small territories that were very dynamic and innovative in the 1980s (Becattini 1992). An industrial district is a geographically concentrated production system that is created through a division of labour between several small, specialized businesses. This cooperation between firms creates the potential for strong *endogenous* growth in the system. Growth is further encouraged when firms cooperate with non-production organizations such as industry associations and financial institutions. It can also rest on the particular skills and knowledge of the local labour force, as was the case in Italian industrial districts.

Successful districts are characterized by a few organizational principles: (i) significant business exchanges between firms and between associations and firms based on a high level of *cooperation* and on strong social cohesion; (ii) the existence of trust relationships that foster the exchange of information and collaboration between firms; (iii) access to a skilled labour pool and opportunities for specialized training; and (iv) the existence of local institutions, either public or private, that monitor industry developments and disseminate information to firms (Klein, Tremblay and Fontan 2003; Klein, Fontan and Tremblay 2001; Tremblay *et al.* 2002, 2003).

Clusters

Clusters, or in Aydalot's (1984) terms, "innovative milieux," are geographic concentrations of firms and supporting organizations that "trust" one another and frequently exchange knowledge. The theory of innovative milieux emphasizes the role of the milieu or cluster as a source of innovation and industry growth: the proximity of competencies promotes the creation of new innovative firms (Aydalot 1984). Our study of Quebec's multimedia industry examines the level of cooperation between local organizations — the key to economic dynamism in this theory. Indeed, the concept of innovative milieux provided the inspiration for the Cité du multimedia project (Tremblay *et al.* 2002). This chapter goes beyond that project to analyze the whole multimedia sector in Montreal and Quebec City.

Local Systems of Production, Interaction and Innovation

Theories of industrial districts, clusters, and local systems of production highlight the importance of formal and informal relations for circulating information, which in turn promotes the development of competencies and innovation. Indeed, information exchanges appear to play an *essential* role in innovation. These theories also suggest that regions (at various geographical levels) are dynamic actors, not passive receptacles of economic activity and that close proximity between actors multiplies their learning capacity and hence their ability to innovate (Veltz 1994).

North American scholars became interested in these European perspectives on industrial districts and innovative milieux when Piore and Sabel published *The Second Industrial Divide: Possibilities for Prosperity* (1984), which emphasized the link between the transformation of productive systems and the socio-economic embeddedness of regions. In recent years, growing interest in the connection between geography and innovation has led to new research into the local region's role in innovation and economic development (Héraud 2003). We now turn to a description of the multimedia sector in Quebec and our preliminary analysis of the sector.

THE MULTIMEDIA SECTOR: SOME DEFINITIONS

Although our primary focus is the Montreal multimedia industry, we often refer to Quebec data, because there is no specific Montreal data available. It must also be noted that there is some multimedia activity outside Montreal and Quebec City where we did interviews.[1] It is also important to note that official statistics on the industry can be misleading because the multimedia and new media sectors are difficult to define and often integrated under the wider heading of information technology industries.

Several previous studies have analyzed the electronic services and multimedia industry (EMS) in Quebec. A study by the Quebec Institute of Statistics published in April 2001 analyses the electronic services and multimedia sector of the province as a whole, as well as specifically for the Greater Montreal Region, Quebec City, and the rest of Quebec's regions. Other studies include TECHNOCompétences's *The Information Technology and Communications Sectors in Quebec—Excluding Manufacturing* (2002), and PricewaterhouseCoopers' *A Study on the Multmedia Sector and Information Highway in Quebec* (2000), commissioned by the Association des producteurs en multimédia du Québec (APMQ),[2] the Centre for Expertise and Service in Multimedia Applications (CESAM), and the Forum des inforoutes et du multimédia (FIM).[3] These offer few specifics on Montreal's industry, however. The most recent studies directly relating to the Montreal area are a series of papers published by Montreal International in 1999 and 2001.[4]

PricewaterhouseCoopers (2000) notes that there are several definitions of multimedia and new media because these are relatively new concepts dating back to the mid-1980s and mid-1990s respectively. It narrows these definitions to the following:

> Multimedia and new media is an industry that develops audio, visual, and textual products and that often places these in an interactive environment. This industry is part of what is referred to as the "new economy."... The multimedia sector therefore includes businesses that offer on-line services (via the Internet or intranet), off-line services via CD-ROM, and interactive services, which can be provided either on- or off-line (PricewaterhouseCoopers 2000, p.i).

The PricewaterhouseCoopers study also quotes the definition of the Canadian Radio Television Council (CRTC) as:

> "New media" designates a diverse range of communication products and services including, but not limited to, video games, CD-ROM, email, online messaging services, fax technology, e-commerce, telecommunications and world wide web and internet services.

We adhere most closely to the definition put forward by CESAM, which corresponds to descriptions offered by industry personnel we interviewed.

> To be considered as "multimedia" a product must use at least three of the four of the following media simultaneously: text, images, sound, and video. Moreover, the product must be interactive. Interactivity is defined as bi-directional communication between the user and the system that is made possible by technology that allows for a rapid response time between the user and the system. In addition, the technology must be easy to understand and use. The words, "new media" and "multimedia" can also be used to refer to the composite materials making up computers, an application system, or software applications (CESAM 1999, p. 6).

A STATISTICAL OVERVIEW OF QUEBEC'S MULTIMEDIA SECTOR

The following overview of Quebec's multimedia sector is based on two sources of statistics. The first is a 2001 study by Institut de la Statistique de Quebec (ISQ) looking at Quebec's electronic and multimedia services sector (EMS). The second source is Quebec government data on the province's information technology industry.[5]

Quebec's Electronic and Multimedia Services Industry

Regional distribution of EMS firms.[6] The following overview of regional distribution of EMS operations divides the province into three geographic areas: the Greater Montreal Region, which is the dominant centre for electronic and multimedia services, the Quebec City region, and all of Quebec's remaining regions combined.

Number of firms. In 1999, there were 3,175 EMS firms: of these, 1,791 (56.4 percent) were specialized and 1,384 (43.6 percent) of which were non-specialized. A firm is specialized if it earns more than 50 percent of its revenues from electronic and multimedia activities and non-specialized if it earns 50 percent or less of its revenues from these activities. Among the specialized businesses, 1,284 provide exclusively EMS products and services. Of total EMS firms in 1999, 2,301 (72.5 percent) were located in the Montreal area and 361 (11.4 percent) were located in the Quebec City area. Approximately 60.4 percent

FIGURE 1
Distribution of Employment by Region

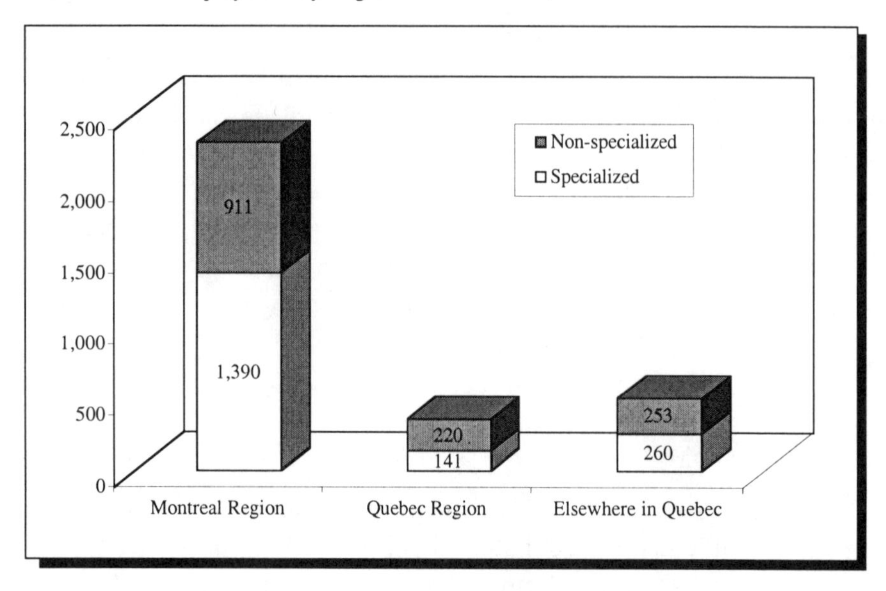

of the Montreal firms were specialized and 39.6 percent were non-specialized. Of the specialized firms, 73.2 percent (1, 018), performed EMS functions exclusively. Approximately 39 percent of the Quebec City firms were specialized and 61 percent were non-specialized. Of the specialized firms, 87.9 percent (124) operated solely in the EMS field. Elsewhere in Quebec in 1999 there were 513 firms in the EMS sector, of which 260 (50.7 percent) were specialized and 253 (49.3 percent) were not specialized.

Areas of activity. In 1999, the production of multimedia content, both on- and off-line, comprised 67.4 percent of EMS activity. Some 51.5 percent of EMS firms provided support services to multimedia production or online services and 35.3 percent offered Internet services. In addition, 27.7 percent of firms offered e-commerce services, teleconferencing, and other electronic services.

Distribution of total sales according to market. In 1999, 79.1 percent of sales in Quebec's EMS industry took place in the provincial market.

Approximately 8 percent of sales were in other Canadian provinces, 7.2 percent of sales transactions were in Europe, and 5.3 percent of sales were in the US.

FIGURE 2
Distribution of Gross Sales by Geographic Market

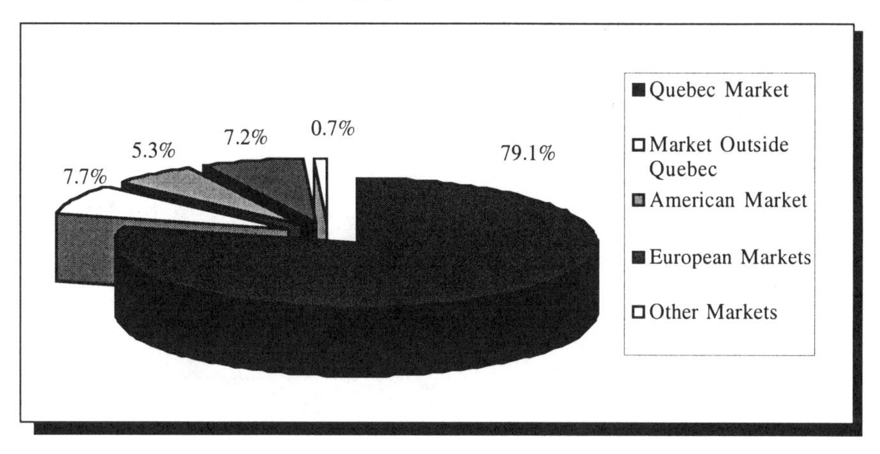

Source: Institut de la Statistique de Québec (2001).

From these data, as well as from interviews, we conclude that there is more interaction between firms in Quebec than between Quebec firms and the international market. Indeed, many firms in both the business and entertainment sectors possess mainly Quebec-based clients. Among the total clientele of Quebec's EMS industry, 68.2 percent were businesses, while other institutions and the public sector comprised 23.7 percent and 6.9 percent of customers respectively.

Distribution of revenues according to market activity. Total revenues in Quebec's EMS industry were $4.097 billion in 1999. This amount can be broken down as follows:

- $1.742 billion for multimedia production.
- $911.1 million for support services for multimedia or online production.
- $672.0 million for Internet services
- $772.4 million for EMS-related activities.

FIGURE 3
Distribution of Revenues by Sector of Activity Within Multimedia and New Media

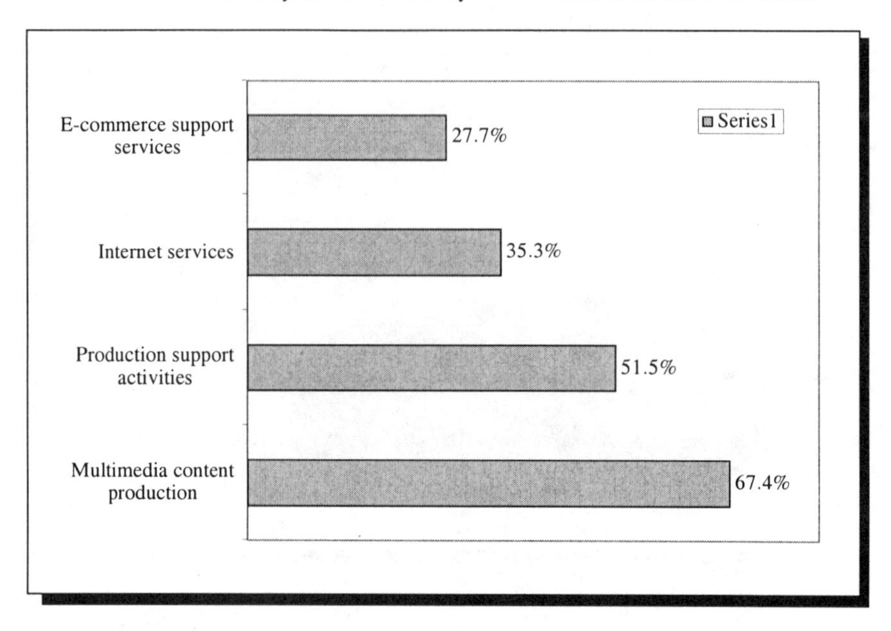

From these numbers, we see that multimedia content production dominates EMS business activities, although this is sometimes intertwined with related services. In our interviews, we tried to take this dimension into account.

Number of employees. In 1999, a total of 63,382 jobs were associated with Quebec's EMS industry. Among these, 24,974, or 39.4 percent, were specialized jobs. Three-quarters of all EMS jobs in Quebec were located in the Montreal area. According to this study, specialized employees possess several skills and work in more than one area within the industry. This statistic does not include autonomous workers, nor does it take into account the fact that many businesses not included in the EMS sector employ personnel to do EMS-related tasks.

FIGURE 4
Number of Specialized and Non-specialized Employees Within Quebec's EMS Industry

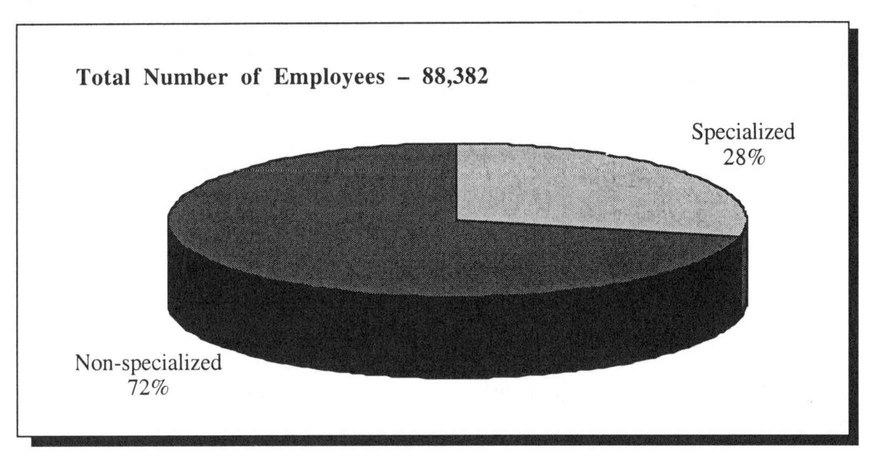

Quebec's Information Technology (IT)

Quebec's information technology industry can be broken down into six subsectors: telecommunications, multimedia, information and software services, e-commerce and electronic media, microelectronics and, finally, computer hardware.[7] A certain number of characteristics make Quebec's IT industry attractive to foreign investors:

- a good balance of innovative local and multinational EMS firms,
- world-renowned creativity,
- a loyal and competent workforce,
- a broadband and high speed optical network,
- a desire to expand to worldwide markets, and
- good research and development institutions.

Investment costs in Quebec's IT sector are considerably more favourable than in the United States (KPMG 2002). For example, in Quebec, investment costs are approximately 12.2 percent more advantageous in electronic components assembly, 24.8 percent in leading-edge software development, 24.6 percent in multimedia content development, and 37.5 percent in the creation and testing of electronic systems (Quebec 2002*a*, based on KPMG 2002). Finally, it is important to remember that 52 percent of Canadian venture capital is managed in Quebec and 61 percent of these funds are invested in technology (Quebec 2002*b*). Below are the principal characteristics of the multimedia industry according to the Quebec government.

Multimedia. The sector's development of special effects software for Hollywood productions such as *Titanic*, *Star Wars*, and *Jurassic Park* speaks to the creativity and leadership of software developers in Quebec. Among the contributions of Quebec software developers to Hollywood projects are 3D animation (Discreet and Softimage), 2D animation (Toonboom), interactive imagery (CAE Electronics), game development (Ubi Soft), multilingual navigation (Alis Technologies), and the development of Internet and e-commerce applications (Copernic and Nurun, a subsidiary of Quebecor, the leading commercial printer in the world, and Public Technologies Multimedia). In 1999 the sector's 1,200 firms employed 12,800 workers and generated revenues of $7.5 million. There are also 32 colleges and eight universities in the region offering multimedia programs. Locating in the Cité du multimédia allows firms to claim a tax credit worth 40 percent of the salaries of eligible employees to an annual maximum of $15,000 per employee.[8] Quebec's multimedia sector also includes a software and information technology industry, which we will discuss here because of our commitment to study all firms situated between the two poles of multimedia and IT. In 1999 the 3,300 firms in this sector employed 26,200 specialized workers and generated $3.5 billion in revenues. Some of these firms, such as CGI, DMR, IBM and Cognicase, are world leaders in computer engineering, specifically in the development and operationalization of computer systems.

The sector also includes software developers, SAP Labs, Oracle and J.D. Edwards, as well as animation and simulation firms, Toonboom, Softimage, Discreet, and Virtual Prototypes. There is one firm specializing in multilingual navigation, Alis Technologies, and two specializing in Internet and intranet development, Eicon Technology and Locus Dialogue.

E-business and electronic media in Quebec. Quebec's e-business and electronic media industries contain world-class service providers, such as BCE Emergis, SAP Labs, CGI, DMR, Cognicase, and IBM, and specialize in adapting business models to the new information technology environment. There is also an e-commerce incentive program in Montreal where firms could benefit from a tax credit for 25 percent of eligible employee salaries until January

2011 (with a ceiling of $10,000 per employee, per year); this program was also abolished in 2003, but firms already signed up maintain their advantages, while no new firms are admitted. The industry has launched a number of important innovations. For example, Public Technologies Multimedia developed the "Virtual Mannequin," which is transforming the way customers make purchases at JC Penney, Les Galeries Lafayette, Land's End, and other clothing vendors. Nurun, a subsidiary of Quebecor, the world's largest print media services company, is responsible for sourcing, commercializing, and managing clothing sales on the Internet for Sears Canada and for the international airline consortium, Star Alliance. The industry also has several large Canadian Internet portals — Canoe (Quebecor Nouveaux Médias), Sympatico (Sympatico-Lycos), Infinit.com (Netgraphe) and Branchez-Vous! (Invention Media). There is also an excellent network of high value-added distribution centres and some 2,000 call centres, two advantages for any electronic retail business. The table below shows revenues, the number of firms, and employment levels in the IT sector in Quebec in 1999, before the dot-com bust and following reductions in personnel.

TABLE 1
Revenues, Firms and Employment in Quebec's IT Sector, 1999

	Revenues $ Millions (1999)	*Firms*	*Jobs*
Telecommunications	13.1	384	38,500
Multimedia	7.5	1,200	12,800
Computers and software	3.5	3,300	26,200
E-business and e-media	n/a	n/a	n/a
Microelectronics	5.0	84	9,850
Computer hardware	1.3	144	7,500

Source: Quebec (2002*b*).

Employment in Montreal's Information Technology and New Media

The most recent study on employment in these sectors was published by Montreal International in 2000.[9] The sector has 61,000 jobs in 1,700 firms. If we include the telecommunications sector this jumps to more than 110,000 jobs in 2,500 firms. The fifteenth largest metropolitan area in North America, Montreal was ranked ninth in 2001 for information technology employment. On a per capita basis, Montreal ranks fourth behind San Francisco/Silicon Valley, Boston, and Dallas. The table below provides an overview of employment levels (including contract positions) of some of the largest multimedia employers in the Montreal region as of July 2001.

TABLE 2
Employment Statistics for Some Large Multimedia Firms in the Montreal Area, July 2001

1,000 employees and more	CAE	4,500	Systèmes électroniques Matrox	1,500
	IBM Canada	2,030	Insight Canada	1,500
	Groupe CGI	1,700	Viasystems Canada	1,400
Between 999 and 500 employees	STS Systems	950	Médisolution	510
	Primetech Electroniques	700	Systèmes SCI	500
	Les Industries C-MAC	540	–	
	Ubi Soft Divertissement	400	Teknor Applicom	250
Between 249 and 130 employees	Gemplus Canada	240	IKon Solutions de Bureau	130
	Softimage Avid	230	Invensys Foxboro	130
	Neexlink	225	Logibro	130
Between 129 and 100 employees	La Société d'informatique Oracle du Québec	120	ZAQ Technologies	100
	Vision Globale	120	Artificial Mind & Movement	100
	Guillemot	120	Summit Technologies	100

Source: Montréal International (2001*b*, Annex B: List of largest firms).

Employment structure in the IT and New Media industries. In 2001, 25 percent of the industry's employment was in the manufacturing sector, which accounted for 15,000 employees in more than 170 firms. CAE Inc. is the principal employer, with others including Systèmes électroniques Matrox, Viasystems Canada, Primetech Électroniques, les Industries C-MAC, Systèmes SCI, and Technicolor. These firms are primarily involved with the manufacture of computers, peripheral devices, and electronic components. Approximately 56 percent of employment is in the development sector, which accounted for 34,300 employees in more than 930 firms. IBM Canada and Groupe Cognicase are the main employers in this sector. The others are le Groupe CGI, le Groupe Conseil DMR, le Groupe LGS, Insight Canada, STS Systems, MédiSolution, Ciné Groupe, Ubi Soft Divertissement, SAP Canada, BCE Emergis, Discreet, Technologies Interactives Médiagrif, Softimage, Artificial Mind & Movement, Kaydara and Toon Boom Technologies. Activities include the development and implementation of software systems, software development, and the digitization and animation of information. The service sector accounts for 19 percent of employment, 11,700 jobs in 600 companies. RBA is the principal employer in this sector. The others are Xerox, Canon Canada, Hartco, and Hewlett-Packard Canada. Activities include distribution, maintenance, Internet access and design, computer development labs.

Institutions of networking, development, and research. Several institutions support the development of multimedia and new media firms. Two of the most prominent are Alliance numériQC and the Centre for the Promotion of Software (Centre de promotion du logiciel). Alliance numériQC's mandate is to support and foster the growth and competitiveness of Quebec's multimedia and information highway; some firms view this organization positively, while others consider it covers too many mandates. The Centre for the Promotion of Software has a mandate to assist and support the growth of Quebec's software firms in local and external markets.

Research centres and specialized organizations. Quebec has several research centres and specialized organizations designed to support Quebec's IT and multimedia industries. Among these are:

- The Montreal Centre for Computer Research (Le Centre de recherche informatique de Montréal) (CRIM). Recognized in Canada and around the world, the CRIM is a centre of excellence in IT development. It conducts R&D activities and oversees the transfer of technologies in software engineering, information-processing systems, multimeda, computer vision, computerized industrial processes, and information communication networks.

- The National Optics Institute (L'Institut national d'optique) (INO). This national centre of optical and photonic research participates in diverse R&D projects in the information technology sector. This organization is the sole manufacturer of specialized fibre optics in Canada.
- Le Centre d'expertise et de services en applications multimédias (CESAM), is a centre that specialized in multimedia technology and applications. The centre no longer exists; it has joined with others to form the Alliance NumériQC.
- The Francophone Centre for the Computerization of Organizations (Le Centre francophone d'informatisation des organisations) (CEFRIO). This research centre conducts research on the use of IT products, strategic planning, and knowledge transfer in firms, the public sector, and in information management.

It is important to note that our interviews did not indicate important relationships between these institutions and the multimedia sector. Aside from some of the ties between certain IT firms and the CRIM, there are virtually no links between these support institutions and IT firms in Quebec. The few relationships that exist are characterized as "superficial," as they usually involve applications for the approval of products in response to government regulations. How can one explain the lack of interaction between public institutions and IT firms in the province?

In response, it is necessary to highlight certain traits of the Montreal multimedia industry. In particular, one must consider two factors. The first is "contextual" and pertains to the environment in which a firm develops. The second set of factors is "structural" and pertains to industry's internal dynamics. Our account of how these factors interact draws on our interviews with actors in the IT sector.

When we refer to "contextual" explanations for the lack of interaction between public and private associational institutions and firms in the multimedia industry, we are referring to the precarious situation of the sector following the downturn of the technology industry in 2000, both in terms of the financial position of individual firms as well as the negative perceptions of institutional lenders. Firms we interviewed are of the opinion that the current climate is not conducive to R&D spending, as research expenditures are difficult to justify when firms find themselves in a position of under-production. The market slowdown and financial difficulties resulting from the downturn are impeding the development of research and development in the sector. Nevertheless, the main reasons for the dearth of ties between public institutions and IT firms relate to the multi-faceted structure of the industry in Quebec.

First, the multimedia industry in Quebec is very young. According to the firms we interviewed there is a correlation between a firm's age and its relationships with research centres. The older firms had previously worked with research centres on short-term or project work. Once the project in question was completed, whether successfully or not, the links between firm and research centre collapsed. Inversely, firms established more recently have rarely worked with research centres. Instead, they concentrate on daily management issues, an activity with which they are increasingly preoccupied given the current context. The small size and limited resources of many of these firms make it difficult to access research centres. Firms often view such access as an onerous expense they cannot afford.

Second, Quebec's centres of IT and multimedia research tend to be undervalued by many multimedia firms. While many firms know the centres exist, they perceive them as being principally geared to large enterprises and do not see the value they can provide to small organizations. Most of the equipment that the multimedia sector uses is not as costly as that used by other new economy industries such as biotechnology. While computer technology becomes rapidly obsolete, it is relatively inexpensive for firms to replace outdated machines. Rather, the acquisition of production software constitutes a more onerous cost for these businesses.

Finally, it is important to note the temporal trajectory along which these firms evolve. This is a sector in perpetual movement, which means that the firms do not have the means to undertake long-term projects.

Training. There are several public and private educational institutions that offer training programs in IT and new media technology at the university, pre-university, and college levels across Quebec. In Montreal alone, these include: eight institutions offering university-level programs; 15 institutions offering pre-university training; and 34 private technical colleges. As with the research centres, relationships between training institutes and firms are feeble. However, links are being established between teaching institutions and new media firms, as firms begin to view educational establishments as the principal producers of labour for the industry. Among these emerging links are internship programs that place new graduates in multimedia firms. None of the 31 firms interviewed has research agreements with training and educational institutions. The few relationships between firms and educational institutions centred on e-learning initiatives.

While links between public institutions of learning and firms are sparse, there are much stronger ties established between the industry and private training institutions. This is explained by the fact that teachers at these institutes work in the private sector. This overlap of personnel has a downside, however, in that private learning institutes cannot offer services to firms for fear of conflicts of interest. In certain cases, the administrative boards of these institutions are replete with personnel from the multimedia sector. Public colleges suffer from a bad reputation among multimedia firms in Quebec. According to many, the proliferation of programs developed in response to the rapidly growing multimedia sector has diminished the quality of teaching in these institutions. A badly trained teaching staff could translate into poorly trained graduates. In addition, there is a fear of producing more graduates than the industry can absorb.

The rapid changes engulfing the multimedia industry in Quebec are a handicap in the establishment of strong links with universities, particularly because universities are perceived as being slow to adapt their programs to benefit the industry. There are signs of some links developing between learning institutions and multimedia firms in Quebec. While these links do not pertain to the production of new products, they do tend toward the creation of a long-term environment that is favourable to the industry. An example of this is Hexagram, which coordinates activities between Université du Quebec in Montreal (UQAM), Concordia, and multimedia firms.

Support for businesses. IT and new media firms benefit from a number of fiscal advantages and public and private financial assistance measures. The Liberal government in Quebec, elected in April 2003, abolished the employment incentives that supported the creation of jobs in designated areas, namely the Cité du multimédia and the Cité du commerce electronique. Firms that had previously signed up for these employment incentives continue to benefit from the program but no new agreements will be made under the now-defunct employment incentive system (Tremblay *et al.* 2002).

Many of the officials we interviewed believe that these forms of public support played a fundamental role in the development of the multimedia industry in Montreal. Some officials wondered whether the industry, especially service-based firms, could survive in the absence of these programs, while others consider that R&D credits are more important. Others think that this will permit us to see whether the advantages linked to the quality and relatively low price of Montreal labour are truly competitive and if they can compensate for the loss of financial incentives.

Venture capital. Studies of industrial districts, such as those in clusters, often refer to the important role of actors that finance firms in promoting innovation. Many venture capital firms specializing in multimedia are located in Montreal. Appendix II lists 40 of the most important firms. Currently it is difficult to acquire venture capital in the multimedia industry in Quebec. According to the venture capital firms we interviewed, requests for capital from e-businesses are usually not entertained. Even to secure an interview with venture capitalists, multimedia firms must present a solid business plan. Most venture capital firms negatively affected by the technology downturn invested strongly in multimedia. In fact, in the late 1990s the industry appeared to be the new Eldorado for investors seeking a quick return. Of course, this heavy investment in technology is obviously not unique to the Montreal area. During our discussions with industry actors, certain individuals recalled a period at the end of the 1990s when it was easy to get financing for multimedia projects even in the absence of viable business plans. Moreover, it was possible for multimedia firms in the 1990s to attract more financing than was requested of venture capitalists.

It is necessary to recognize that because multimedia was an emerging sector at the end of the 1990s, business plans were highly experimental. It is also difficult for investors to understand the subtleties and complexities of the sector. In fact, most investments were in "knowledge capital" rather than tangible goods. Investors were therefore left with nothing when firms declared bankruptcy. The industry is also recognized for underinvestment by its firms. In light of this, it is obvious that the sector has greatly evolved in a short period of time and there are at present few ties between the multimedia sector and venture capital.

General Characteristics of the Multimedia Employment Pool in the Province of Quebec and Montreal

Theories of industrial districts and clusters highlight the importance of the workforce as a factor of attraction for firms and as a source of innovation in a given sector or region.

Employment. The first salary study done in the multimedia industry was conducted in 2000 by TECHNO*Compétences* and the Quebec Centre for the Promotion of Software (Centre de promotion du logiciel québécois).[10] This study focused on 39 types of jobs associated with the software and multimedia industries. The study revealed that:

- The industry attracts mostly masculine employees.
- The main recruitment difficulties are the lack of qualified candidates (75 percent), competition between firms (40 percent), and the salary expectations of candidates (32 percent).
- The most sought-after skills are programmer-analyst, software development, network administration, project management, and technical writing.
- The highest salaries in the multimedia industry are those of project managers ($58,588 on average), project integrators ($57,690 on average), and artistic directors ($49,158 on average).
- Of the 15 most sought-after positions, 13 require university-level training.

Types of jobs. Since activities in the multimedia industry are widely diversified, employment requirements differ enormously by job type. Robitaille and Roy (1999) break down the multimedia labour force into three profiles: interactive support specialists, computer specialists (programming or content integration) and, finally, content specialists. This description appeared to hold during interviews in the field.

Modes of work. Unlike the general labour market, the forms of employment are highly diversified in the multimedia sector. Elsewhere, we have described the mode of work in this field as "nomadic" or "boundaryless" (Tremblay 2002, 2003*a*,*b*,*c*,*d*). We observed here that many young workers of the multimedia sector are self-employed or independent workers, and many are teleworkers, working from home for a given firm (Tremblay 2003*d*). This caused us to link these observations to the "boundaryless careers" approach.

In the boundaryless careers approach, three different components are used to account for career conditions in a knowledge-based economy. Like the resource-based strategy approach, which identifies three components of the firm (core competencies, networks of partnerships, and organizational identity), the boundaryless careers approach makes the distinction between three components of competency which are continuously interacting with each other:

(i) *Knowing how*, which refers to knowledge, abilities and even acquired routines; (ii) *Knowing whom*, which implies privileged relationships, social networks and contacts; and (iii) *Knowing why*, which takes into account interests, passions, values and identity construction (Cadin *et al.* 2000, cited in Tremblay 2003*a*). We found the three elements present in the multimedia sector, but we observed that *knowing whom* or network relations, and *knowing why* or passion and shared interests, played an extremely important role in how multimedia workers got contracts, developed their skills and their career paths. Another important aspect of this approach is that it considers careers on the basis of different social spaces, communities of practices or informal communities, rather than the organizational or internal dimension (Tremblay 2004; Sharp 1997). This is important in this sector because workers rely on skills and competencies developed by friends and colleagues working for other firms. Thus, the informal network or community of practice is extremely important. We would add that network relations appear to be more important between individual workers than between firms per se.

In the multimedia sector informal communities help individuals share information, ways of seeing things, ways of thinking, tricks of the trade or job opportunities (Cadin *et al.* 2000; Tremblay 2002, 2003*a*). These types of exchanges have been observed in studies on business networks, in particular the case of the Silicon Valley (Cadin *et al.*, 2000). It is precisely this type of network and exchanges that we observed in our interviews with multimedia workers (Tremblay 2002, 2003*a*,*b*). We also observed that human resources management strategies are rather absent altogether; employees as well as employers appear to prefer a more informal type of relations within the firm, and the semi-autonomous teamwork we might expect to find in such a sector is therefore absent, or at least not at all formalized (Tremblay and Rolland 1998).

This concept of a boundaryless career, of skills development through mobility, is the main point that we observed in our interviews with workers in the multimedia sector. The mode of career development observed in the sector can thus be represented by Figure 5, inspired by Cadin *et al.* (2000; for more detail, see Tremblay 2002, 2003*a*,*b*,*c*).

In this context, employees move frequently between firms (or at least they did when the sector was booming). The workers are set in institutional contexts where they develop expertise and career competencies through their work experience, which permits the development of what can be considered a

FIGURE 5
Skills Development in Boundaryless Careers

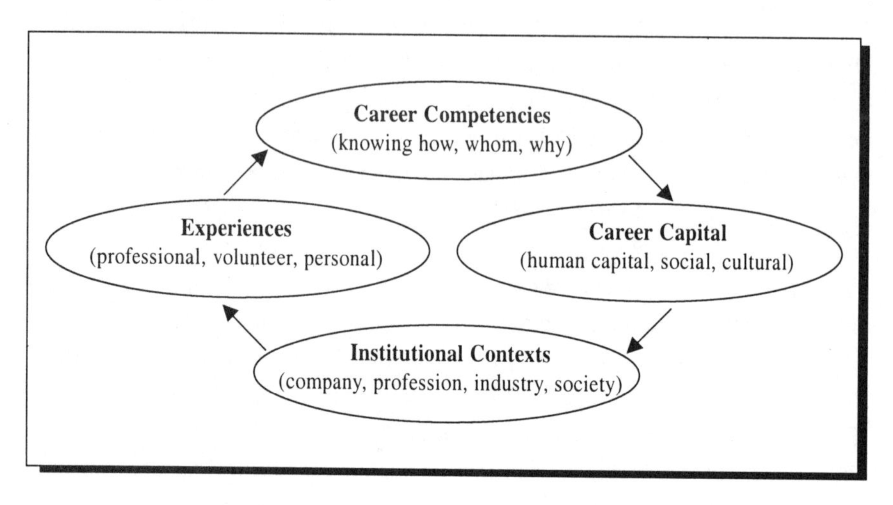

career capital. As suggested by Nonaka and Takeuchi (1995), experience is considered not only in terms of assets but also in knowledge management. These authors show that creation of knowledge requires successive phases of integration of knowledge, clarification of tacit elements, then reinternalization of the explanation (Tremblay 2001; Tebourbi 2000; Cadin *et al.* 2000).

Career capital results from an accumulation of career competencies, as defined above. Economists are familiar with the concept of human capital (Tremblay 1997; Tremblay and Rolland 1998), but nomadic career theorists view the concept in its broader sense. Many scholars use the metaphor of the competency portfolio to underline the freedom of the actor, and the fact that he or she is continuously making choices, whether or not they are aware of it. These choices are then translated into "competency effects" which are derived from real-life experiences (Tebourbi 2000; Cadin *et al.* 2000). It is thus a circular flow originating from experience to form career competencies, which create career capital, and which will only make sense in a given institutional context. This is very much in line with what is observed in our interviews with workers in the multimedia and new media sectors (Tremblay 2002, 2003*a*,*c*).

As we observed, it is mainly in employment that learning and training take place in multimedia and new media; although there is specialized training on software offered by firms — by external trainers, or specialized institutions — there are few relations between multimedia firms and educational institutions, since firms consider that college and university programs do not follow technological evolution.

CONCLUSION-ANALYSIS

This chapter reports an ongoing research, and more thorough analysis of the interviews will be carried out in the coming months. Here we limit ourselves to a few insights drawn from the research to date and try to relate them to theory.

In research and writings on clusters, districts, and the role of territories in innovation, authors often note the spontaneous nature of interaction between firms, but some also indicate that governmental or local institutions have been tempted to replicate these interactions. In our case, we find that there was some level of spontaneous gathering of firms in the centre of Montreal, but the importance of supporting institutions and financial benefits has also attracted some to the sector and more particularly to the Cité du multimédia, where employment subsidies were offered.

Concerning the question of whether the multimedia firms in Montreal constitute a cluster, we can suggest that some elements of clusters are present here, but it is still unclear if this constitutes a cluster. Clusters or districts usually have a certain history behind them, while multimedia is a new sector of activity and it is presently difficult to determine if the grouping of multimedia firms in Montreal will endure over some decades, especially since the employment incentives have now disappeared in Quebec. Multimedia is part of a larger IT sector and this sector has rapidly developed over recent years in comparison with other sectors. It is clear that the life cycle of the product and industry will continue to evolve, which raises the question of what the multimedia industry will look like in the future. Of course, these questions are due to the fact that we are studying a sector or cluster that is in its infancy when compared to sectors such as aeronautics, automobiles or others in Canada.

We also observe that the sector is characterized by a limited degree of collaboration between firms and between firms and other sector actors. However, a more detailed analysis is needed to confirm the nature and extent of this collaboration and how these collaborations impact the industry. Clearly the mobility of workers between firms plays a role in the mobility of information and knowledge in the sector and possibly the cluster, but this also remains to be studied in more detail. We must determine if there are linkages between firms, which do not appear to be very important to date, except for a few cases. Most firms offer services or produce their products with their own workers and sometimes self-employed workers (less than we had expected, however), but there does not seem to be a very strong local value chain in the multimedia sector per se, contrary to other sectors in the IT industry. We noted that some firms had to some extent acted as incubators for others, with workers learning skills in these firms and then moving on to create their own businesses. But again, with the difficulties known in the sector in recent years, this type of incubation and spinoff has tended to diminish.

A definite conclusion on cluster characteristics and the qualification of the Montreal multimedia industry as a localized system of production or district requires a more detailed analysis of our data and a few more interviews with specialists in the sector, especially since provincial government support has diminished in the last year. While more general R&D support and fiscal incentives remain, the strong employment incentive offered to multimedia firms which established themselves in the Cité du multimédia was abandoned in 2003 by the Liberal Party. This will cause analysts of the sector to evaluate whether the employment incentives contributed largely to the expansion of the sector, or whether other factors still come into play, as some contend (creativity, connection with education, low costs in comparison with the US).

It is clear that firms have not yet developed strong relations between themselves and with institutional actors, but again, the sector is very young and radical changes in government support, as well as recent difficulties in the IT industry, may contribute to a new need for solidarity and an emergence of new initiatives of collaboration and exchange. This remains to be seen, and the interviews to be done in the next year will permit a more complete evaluation of the character of the sector, as well as its possible characterization as a district, cluster or localized system of production. Indeed, it is clear that the geographical proximity between actors can favour some form of territorial construction, cluster or localized system of production. However, as we have observed elsewhere (Tremblay *et al.* 2003), for this to be the case, proximity needs to be not only geographical but also relational or organizational, that is, based on a sharing of experience, routines, representations, rules or codes on which coordination can be built. While this sharing of experience and routines does exist between the workers of the sectors (Tremblay 2002, 2003*a*,*b*), it does not yet exist between firms and other institutional actors. We must add that in all forms of territorial construction, history plays an important role and it is clear in the case of the multimedia sector that history has not yet had enough time to play out into a process of path dependency that would lead to a process of clustering.

NOTES

Diane-Gabrielle Tremblay is Canada Research Chair on the Socio-organizational Challenges of the Knowledge Economy; Catherine Chevrier and Serge Rousseau are research assistants at Télé-université.

[1]The case is also partially documented in Simard (2003).

[2]Association des producteurs en multimédia du Québec (Quebec Association of Multimedia Producers).

[3]Forum des inforoutes et du multimédia (Information Highway and Multimedia Forum).

[4]The first is entitled *L'industrie des technologies de l'information et du multimédia dans la région de Montréal* (1999) and the second, *L'industrie des technologies de l'information et des nouveaux médias du Grand Montréal* (2001). There is also a study entitled, *Analyse de l'industrie du multimédia à Montréal* (1999) prepared by Robitaille and Roy in the context of MA studies at the University of Montreal directed by Claude Manzagol.

[5]The statistics herein date to 1999.

[6]Based on Institut de la Statistique de Québec.

[7]Statistics and information in this section are drawn from the Web site, http://www.infostat.gouv.qc.ca/iq/, Gouvernement du Québec (2002) *Investing In Quebec* (Investir au Québec). Unfortunately, this information dates to 1999 and there has been no more recent study of Quebec's IT sector since the downturn of 2002. Thus, certain numbers should be lower, as indicated by our interviews with firms in 2002–2003.

[8]Note that in the spring of 2003, the Liberal government in Quebec abolished the employment incentives previously available to designated high-tech industries. All existing agreements will be honoured, but no new firms will be able to benefit from this subsidy.

[9]Statistics in this part of the chapter are drawn from a report by Montréal International (2001*a,b*).

[10]Since renamed the Réseau Inter logiq.

REFERENCES

Aydalot, P. 1984. "A la recherche des nouveaux dynamismes spatiaux," in *Crise et espace*, ed. P. Aydalot. Paris: Economica.

Becattini, G. 1992. "Le district marshallien: une notion socio-économique," in *Les régions qui gagnent. Districts et réseaux — les nouveaux paradigmes de la géographie économique*, ed. G. Benko and A. Lipietz. Paris: Presses universitaires de France.

Cadin, L. A.F. Bender, V. Saint-Giniez and J. Pringle. 2000. "Carrières nomades et contextes nationaux," *Revue de gestion des ressources humaines*. Paris: AGRH, pp. 76-96.

Centre for Expertise and Service in Multimedia Applications (CESAM). 1999. *Mémoire au CRTC sur les nouveaux médias 1999*. Montreal: CESAM.

Fontan, J.-M., J.-L. Klein and D.-G. Tremblay. 2001. "Mobilisation communautaire et gouvernance locale: le technopôle Angus," *Politique et Sociétés* 20(2/3):69-88.

Héraud, J.-A. 2003. "Régions et innovations," in *Encyclopédie de l'innovation*, ed. P. Mustar and H. Penan. Paris: Economica.

Interactive Multimedia Producers Association of Canada. 2001. *Profile of the Canadian Multimedia Industry.*

Institut de la Statistique de Québec. 2001. *Rapport d'enquête sur l'industrie québécoise des services électroniques et du multimédia 1999.* Quebec City.

Institut de la Statistique de Québec. Various years. *S@voir.stat, Bulletin sur L'économie du savoir.*

Klein, J.-L., D.-G. Tremblay and J.-M. Fontan. 2003. "Systèmes locaux et réseaux productifs dans la reconversion économique: le cas de Montréal," *Géographie, économie et société* 5(1):59-75.

Klein, J.-L., J.-M. Fontan and D.-G. Tremblay. 1999. "Economic Reconversion, Partnership and Community-based Mobilization in Montréal: Towards the Activation of Socio-territorial Capital," *Zeitschrift fur Kanada-Studien* (March):120-37.

—— 2001. "Les mouvements sociaux dans le développement local À Montréal : deux cas de reconversion industrielle," *Géographie Économie Société* 3(2):247-80.

KPMG. 2002. *Comparing Business Costs in North America, Europe and Japan*. Montreal: KPMG.

Montréal International. 2001*a*. *L'industrie des technologies de l'information et du multimédia dans la région de Montréal - 1999.* Profil industriel. Available at <www.montrealinternational.com>.

—— 2001*b*. *L'industrie des technologies de l'information et des nouveaux médias dans la région de Montréal – 2001.* Profil Industriel. Available at <www.montrealinternational.com>.

Montréal Techno Vision. 2001. *Indicateurs de performance 2000*, Montréal et sa région.

Nonaka, I. And H. Takeuchi. 1995. *The Knowledge Creating Company: How Japanese Companies Create the Dynamics of Innovation.* New York: Oxford University Press.

Piore, M. and C. Sabel. 1984. *The Second Industrial Divide: Possibilities for Prosperity.* New York: Basic Books.

PricewaterhouseCoopers. 2000. *Étude du secteur du multimédia et des inforoutes au Québec (A Study on the Multimedia Sector and Information Highway in Quebec).* Montreal: PricewaterhouseCoopers.

Quebec. 2002*a*. At <http://www.invest-quebec.com/fr/int/secteur/secteurs.jsp?nodeid=12018186>.

Quebec. 2002*b*. *Investir au Québec/(Investing in Quebec)*. Quebec City. At <http://www.infostat.gouv.qc.ca/iq/>.

Robitaille, E. and P. Roy. 1999. *Analyse de l'industrie du multimédia : le cas de Montréal*, travail dirigé de M.Sc. sous la direction de C. Manzagol (avec la collaboration de J.-F. Cappuccilli, Mélanie Robertson et Marcel Sauvé), Département de géographie, Université de Montréal. At <http://www3.sympatico.ca/eranlo.rob/multimedia.htm>.

Sharp, J. 1997. "Communities of Practice: A Review of the Literature." Unpublished paper.

Simard, C. 2003. "La contribution des technologies de l'information à la revitalisation du centre-ville de Québec," in *La nouvelle économie: où? quoi? comment?* ed. D.-G. Tremblay. Québec: Presses de l'université du Québec.

Tebourbi, N. 2000. "L'apprentissage organisationnel. Note de recherche de la Chaire Bell sur les technologies et l'organisation du travail." Available on <www.teluq.uquebec.ca/chairebell>.

TECHNO*Compétences*. 2000. *L'enquête Salariale: Enquête 2000 sur la rémunération dans le secteur du logiciel et du multimédia*. 12 August. Available at <www.technocompetences.gc.ca.pdf>.

—— 2002. *The Information Technology and Communications Sectors in Quebec: Excluding Manufacturing (Secteur des services et des technologies de l'information et des communications (TIC) au Québec – excluant la fabrication)*. Montreal: TECHNOCompétences. At <www.technocompetences.qc.ca>.

Tremblay, D.-G. 1997. *Économie du travail: les réalités et les approches théoriques*. Montréal: Editions St-Martin.

—— 2001. *L'innovation continue*. Montréal: Télé-université. Document de base et recueil de textes pour le cours RIN 4120.

—— 2002. "Informal Learning Communities in the Knowledge Economy," in *Proceedings of the 2002 World Computer Congress*. Montreal: Elsevier Press.

—— 2003*a*. "New Types of Careers in the Knowledge Economy? Networks and Boundaryless Jobs as a Career Strategy in the ICT and Multimedia Sector," in *Communications & Strategies*. Montpellier-Manchester: IDATE.

—— 2003*b*. *New Types of careers in the Knowledge Economy? Networks and Boundaryless Jobs as a Career Strategy in the ICT and Multimedia Sector*. Research Note of the Canada Research Chair on the Knowledge Economy No. 2003-12A. Available at <http://www.teluq.uquebec.ca/chaireecosavoir>.

—— 2003*c*. "Nouvelles carrières nomades et défis du marché du travail: Une étude du secteur multimedia," *Carriérologie* 9(2):254-80.

—— 2003*d*. "Telework: A New Mode of Gendered Segmentation? Results from a Study in Canada," *Canadian Journal of Communication* 28(4):461-78.

—— 2004. "Virtual Communities of Practice: What Impacts for Individuals and Organizations?" Forthcoming in the *National Business and Economics Society 2004 Conference Proceedings*.

Tremblay, D.-G. and D. Rolland. 1998. *Gestion des ressources humaines; typologies et comparaisons internationales*. Québec: Presses de l'université du Québec.

Tremblay, D.-G., J.-L. Klein, J.-M. Fontan and S. Rousseau. 2003. "Proximité territoriale et innovation: une enquête sur la région de Montréal," *Revue d'économie régionale et urbaine* No. 5-2003. Paris: CNRS.

Tremblay, D.-G., J.-M. Fontan, J.-L. Klein and D. Bordeleau. 2002. "The Development of the Relational Firm: The Case of the Multimedia City in Montréal," in *Knowledge, Clusters and Regional Innovation: Economic Development in Canada*, ed. A. Holbrook and D. Wolfe. Montreal and Kingston: School of Policy Studies, Queen's University and McGill Queen's University Press.

Veltz, P. 1994. *Des territoires pour apprendre et innover*. Paris: Edition de l'Aube.

APPENDIX I
SUPPORT FOR THE MULTIMEDIA SECTOR IN THE MONTREAL REGION

Fiscal Support

Incitatifs à l'emploi

Centre de développement des technologies de l'information (CDTI)
Cité du commerce électronique
Cité du multimédia à Montréal
Carrefours de la nouvelle économie

Incitatifs à l'investissement

Congé fiscal à l'égard de projets majeurs d'investissements
Amortissement accéléré et congé de taxe sur le capital
Congé fiscal pour les nouvelles sociétés

Incitatifs à la R&D

Déduction de toutes les dépenses admissibles de R&D
Crédit d'impôt de base de 20 percent
Crédit d'impôt remboursable de 20 percent sur les salaires
Crédit d'impôt remboursable de 40 percent
Aide fiscale bonifiée
Congé fiscal pour les chercheurs étrangers

Government Assistance

Programme d'amélioration des compétences en sciences et technologie (PACST)
Fonds pour l'accroissement de l'investissement privé et la relance de l'emploi (FAIRE)
Programme Garantie PME
Programme Déclic PME
Programme d'aide à la recherche industrielle (PARI)
Partenariat Technologique Canada (PTC)
Programme Idée – PME
Programme Micro-Entreprises

Programme de financement des petites entreprises du Canada (FPEC)
Fonds de l'autoroute de l'information
Programme d'aide au développement des communications (PADEC)
Programme Nouvelle Économie
Fonds d'expérimentation en Multimédia
Programme de crédit d'impôt pour la production de titres multimédias
Programme de coopération industrielle de Montréal (PROCIM)
Fonds national de formation de la main-d'œuvre (FNFMO)
Impact PME – Volet Innovation
Programme d'aide à la concrétisation de projets industriels (PACPI)
Programme de soutien aux vitrines technologiques (PSVT)
Programme de soutien au développement de la culture scientifique et technique
Initiative pour contrer les pénuries de main-d'œuvre dans les technologies de l'information

APPENDIX II
VENTURE CAPITAL FIRMS IN THE MONTREAL REGION

Venture Capital Firms in the Montreal Region

Accès Capital, réseau régional de la CDPQ, Argo Global Capital, Banque CIBC, Banque de développement du Canada-Laval, Banque de développement du Canada-Longueuil, Banque de développement du Canada-Montréal -Place Ville Marie, Banque de développement du Canada-Montréal – rue Sherbrooke, Banque de développement du Canada-Ville St-Laurent, Benvest Capital, BMO Nesbitt Burns, Capimont, Capital CDPQ, Capital Communication CDP, Capital International CDPQ, CDP Sofinov, Corporation financière Netzkapital, Fondaction, Fonds de développement emploi-Montréal, Fonds d'investissement de la culture et des communications, Fonds régional de solidarité Ile de Montréal, S.E.C., Fonds régional de solidarité Laval, S.E.C. , Fonds régional de solidarité Montérégie, S.E.C., Gestion Miriva, GTI Capital, Hydro-Québec Capitech, Innovatech du Grand Montréal, Investissement Desjardins, Investissement Québec, Investissements Blue Ship, Investissements Novacap, La corporation placements Banque Royale, Les placements Telsoft, Miralta Capital II, Roynat - Banque Scotia, SIPAR - Société d'investissements en participations, Société de développement des entreprises culturelles (SODEC), Société générale de financement du Québec (SGF), T2C2/Info, TechnoCap, Yorkton Securities

APPENDIX III
WEB SITES

Canada

Canada Economic Development:
http://www.dec-ced.gc.ca/fr/menu.htm
Stratégis Canada: http://strategis.ic.gc.ca/
Statistics Canada: http://statcan.ca
Information Technology Association of Canada: http://www.itac.ca
Infoexport Canada: http://www.infoexport.gc.ca/
Réseau Scolaire Canadien: http://www.schoolnet.ca
Banque de développement du Canada: http://www.bdc.ca
Le projet Canarie: http://www.canarie.ca
Patrimoine Canada: http://www.pch.gc.ca
Téléfilm Canada: http://www.telefilm.gc.ca
Internet news & Ressources: http://Canada.internet.com

Quebec

APMQ: www.apmq.org
Cefrio, Centre francophone d'informatisation des organisations: http://www.infometre.cefrio.qc.ca
Centre de recherche information de Montréal: http://www.crim.ca/
Centre de promotion du logiciel québécois: http://www.cplq.org
Consortium multimédia Césam: http://www.cesam.qc.ca
Forum des inforoutes et du multimédia: http://fim.org
Le marché international du multimédia: http://www.mim.qc.ca
Multimédia-Québec (portail sur l'industrie): http://www.multimedia quebec.com
TECHNOCompétences : www.technocompetences.qc.ca
Montréal International : www.montrealinternational.com

9

Cluster or Whirlwind? The New Media Industry in Vancouver

Richard Smith, Jane McCarthy and Michelle Petrusevich

Introduction

Few forms of economic organization have caused as much controversy as clusters. One might expect firms to distance themselves from competitors, but anyone who has walked through a garment or diamond district or driven through the industrial valleys of Europe knows that this is often not the case. This pattern of agglomeration is pronounced, even setting aside firms that cluster around access to raw materials or power sources.

Silicon Valley, located south of San Francisco on the west coast of the United States, is the common example of a successful cluster of firms, but others abound, ranging from the "Route 128" computer and biotechnology clusters near Boston (Saxenian 1994; Braczyk, Fuchs and Wolf 1999; Lievrouw and Livingstone 2002) to the light manufacturing and textile firms clustered in Italy's Emilia-Romagna region. These and similar cases attract regional governments that would like to replicate the success in their own areas. Regional governments also appreciate the fact that they control many of the policy mechanisms that promote cluster development. Canada's federal government has been significantly influenced by the cluster model since it commissioned a study of the country's economy in the early 1990s. The concept was further entrenched by the more recent work of the Cluster Mapping Project at Harvard (Porter 1991; 1998*a*). The 1991 study replicated for Canada earlier work by Michael Porter on national competitiveness in a variety of developed countries. It identified local or regional cooperation as a factor in the subsequent international competitiveness of nations. Subsequent federal policy introduced several measures to improve competitiveness in Canada, including policies to promote economic clusters. One consequence of such measures has been support for new research on the role of clusters within national systems of innovation.

WHAT IS A CLUSTER?

It is uncertain when firms first perceived the benefits of clustering with other similar firms. The existence of regional reputations for excellence in various industries around the world suggests this pattern has long been part of the economic landscape. The increasing interest in clusters originated in a number of social science disciplines. The earliest modern description goes back to the turn of the nineteenth century when Alfred Marshall (1890) described the phenomenon of industrial districts. The study of clusters also stems from critical analyses of economic organization, particularly the identification of the market-hierarchy spectrum that linked organizational forms and the boundary of the firm to the nature of economic transactions (Coase 1937; Williamson 1975). Relatively uncertain and frequently recurring transactions that required substantial investments were said to be most efficiently carried out within hierarchically organized firms. In contrast, relatively straightforward, non-repetitive transactions not requiring investments were said to take place most efficiently across the market. All other forms of economic organization were said to be hybrids of these two. This framework has shaped much of economic and business thought, but it has also been criticized for ignoring the extent to which economic action is embedded in a cultural and social context. Many historians and sociologists contend that the market is not an amoral self-subsistent institution, but rather, a cultural and social construct (Reddy 1984). Networks such as clusters are now regarded as distinct forms of economic organizations that confer substantial economic advantages, especially in unpredictable and rapidly changing environments. From a management and economics perspective, Porter defines clusters as "geographic concentrations of interconnected companies, specialized suppliers, service providers, firms in related industries, and associated institutions (for example, universities, standards agencies, and trade associations) in particular fields that compete but also cooperate" (1998*a*, p. 78).

Clusters are now more prominent than they have ever been. Porter (1998*a*) claims they are critical to a region's competitive capacity. In the first instance they lead to increased productivity by enabling better access not only to skilled labour, suppliers, institutions, and public goods such as government funding, but also to specialized information, made possible through the development of trusting relationships. When small firms cooperate they are able to increase productivity by leveraging economies of scale that they cannot access individually. Proximity also allows firms to compare themselves directly with competitors. Clusters are innovative milieux within which proximity and trusting relationships facilitate the exchange of knowledge. Moreover, proximity raises the pace of innovation as companies compete to maintain a competitive advantage. Finally, Porter notes that it is not surprising that the dynamic environment of clusters gives rise to many new companies. As suppliers are drawn to a concentration of customers, the barriers to entry may become lower than elsewhere. Porter likens this dynamic to a positive feedback loop, in which the advantages of an expanded cluster benefit all firms within the cluster.

The success of Silicon Valley has spawned a flurry of cluster-formation initiatives (Wolfe and Gertler 2003; Cooke 1999). In a thorough critique of cluster studies, Wolfe and Gertler (2003) identify four characteristics of a cluster: inflows, outflows, local social dynamics, and historical path dependences. *Inflows* are the most readily accessible and measurable of the indicators listed. They identify three different forms: capital, people, and knowledge. *Outflows* are arguably "the best indicator[s] of wider recognition of ... a region" (Wolfe and Gertler 2003, p. 30), as they measure not only goods and services, but also intangibles such as intellectual property. *Local social dynamics* concerns the presence of both collaborating and competing firms, as well as the presence of supporting institutions, public, private, and hybrid. *Historical path dependences* are possibly the best test of "true" cluster dynamics because they assess "the alleged cluster's resilience and robustness over time in the face of severe shocks" (ibid., p. 31), which is one measure of a cluster's ability to adapt. A cluster's success depends on both economic and social dimensions (Piore and Sabel 1984; Pyke and Sengenberger 1992) with emphasis on the social and cultural embeddedness of clusters. Amin claims that "for industrial districts to develop, it is necessary that such a population merge with people who live in the same territory, and who in turn possess the social and cultural features (social values and institutions) appropriate for a bottom-up industrialization process" (2000, p. 14).

Further, Becattini maintains that "the firms become rooted in the territory, and this result cannot be conceptualised independently of its historical development" (1990, p. 40). Successful clusters such as the Italian industrial districts and Silicon Valley cannot truly be understood without reflecting on their cultural and historical dimensions.

Clusters appear to be more closely bound up with dynamic relationships than is the case with other forms of economic organization. Theories provide heuristic relationships between factors and characteristics, rather than black and white normative descriptions. For instance, Porter (1998*b*) explains how proximity between firms enables a cluster to function. Cooke makes the point that many cluster definitions are static, "whereas the key feature of clusters is that they are dynamic" (Cooke 1999). He defines a cluster as "geographically proximate firms in vertical and horizontal relationships, involving a localised enterprise support infrastructure with a shared developmental vision for business growth, based on competition and co-operation in a specific market field" (ibid., p. 292).

A cluster's capacity to adapt may be regarded as the outcome of the interplay between three sets of grouped factors: environmental factors, knowledge factors, and society factors. *Environmental factors*, which relate to the environment in which the cluster is embedded, include the market that the cluster is addressing, demand levels and their relative predictability, the pace of change of technology, numbers of competitors, stage in the technological life cycle, and importantly, the overall size of the market. *Knowledge factors* relate to a cluster's ability and means of access to knowledge. These include the levels of research and development, the levels of exploitation of knowledge, the micro-diversity of cluster players, and the rate at which knowledge decays and is superseded. *Society factors* relate to the ability of actors within the cluster to share knowledge and to coordinate resources, which is often facilitated through the development of trustful relationships. Such factors include the levels of cooperation and competition, the acceptance of common social norms such as reciprocity, the importance of reputation, and the effectiveness of agency. The success and longevity of a cluster depend on the interplay between these factors. Constant change and adjustment are a significant part of this dynamic picture, both at the level of the cluster and the individual actor.

Clusters are also dynamic at further levels of analysis. They are socio-economic phenomena whose effectiveness is linked to a specific technology. As the technology moves through its life cycle, so the cluster changes accordingly. When the technology becomes defunct, or is superseded, the social identity of the cluster may outlive its economic function. It then may serve as a mature start-point for the exploitation of a new technology. Studies of Silicon Valley (Saxenian 1990, 1994) and in the Italian districts (Sengenberger and Pyke 1992; Coro and Grandinetti 1999) show that successful clusters move through a series of different technological waves. They respond to both continuous and discontinuous change. Some regard clusters as a specific stage of development in the process of industrialization (Dimou 1994). In this view a cluster is not stable and static, but continually changing (Garofoli 1991). We can speculate that a cluster evolves in response to its environment in order to survive. Clusters that evolve successfully become more mature and resilient. Most empirical studies of clusters focus on established and successful clusters and relatively little is known about very young clusters. Our current study concerns a cluster so new that it is difficult to say whether it is a cluster at all, or whether it is a temporary, economic phenomenon focused temporarily around British Columbia; in short, whether this is a cluster or a whirlwind effect.

Understanding the New Media Industry

The new media industry is difficult to define. Multimedia, the original term for the industry, made more sense a few years ago when it was easy to distinguish between new and traditional forms of media. While traditional forms still exist they can no longer be separated from new media because they have been strongly affected by the innovations in information and communications technologies. There is scarcely a newspaper without a Web site, a television network, a "streaming media" counterpart, or a publisher without CD-ROM versions of its book lists. This process of merging media forms, referred to as media "convergence," has been extensively studied (Olson 1988; Gilder 1993; Castells 2001).

Defining "new media" is an ongoing challenge. In Canada, the BC government now faces that challenge as it attempts to quantify the industry. BCStats, in conjunction with the Ministry of Competition, Science and Enterprise, issues annual reports, the latest of which is entitled, *Profile of the British Columbia High Technology Sector 2002 Edition*. The report, which attempts to map North American Industry Classification System (NAICS) codes to new media job categories, found that many job functions within new media are either not described by these codes or fall under multiple codes. For example, because new media companies focus on both content and computing their work can be coded as both a product and a service.

To illustrate the variation in definitions of new media, we present the following examples. The Vancouver Public Library, which created a directory of "new media" firms in 2001, used a definition of new media that is similar to ours but has greater emphasis on communication. New media is defined "not so much an industry as a set of applications, software, skills and techniques that are adaptable to a variety of fields. The underlying theme is communications: digital, interactive, wireless, convergent, networked, mobile business and personal communications" (Housser and Vancouver Public Library 2001, p. 3). The Canadian Radio-Television and Telecommunications Commission (CRTC) defines new media as encompassing, singly or in combination, and whether interactive or not, services and products that make use of video, audio, graphics and alpha-numeric text, and involving, along with other more traditional means of distribution, digital delivery over networks interconnected on a local or global scale. Each new publication about the industry seems to introduce a new definition. A new book from MIT press titled *The New Media Reader*, for example, contains eight definitions (Wardrip-Fruin and Montfort 2003).

For this study, we sought to avoid the problem of presenting an overly inclusive definition of new media, but we also wanted to have a definition that was more than a simple list of "approved" technologies or media forms. We argue that the key feature distinguishing a new media firm from one that is merely using new media technologies as part of its business is the extent to which the firm can be said to be innovating in *both* of the two defining aspects of new media, namely "content" and "computation." Without the presence of both of these elements, the potential for new media does not exist.

During our interviews with industry members one interviewee maintained that there is not a new media industry per se, but only a group of traditional media companies integrating new technologies into their existing business models. This raised the question of whether or not new media acts as a new business model, rather than a traditional product or service. Despite this unorthodox view, the interviewee supported our working definition of new media: "firms located in the space between content and computation." This definition helps put a boundary on a traditionally "fuzzy" industry. It also reinforces the notion that there is a value chain in the industry (Porter 1985).

To better understand our definition, it is helpful to think of a continuum (as reflected in Figure 1) that extends from a "pure" content firm (such as a script writer), to a pure "computation" firm (such as a designer of microelectronics). These extremes of the new media spectrum would be outside our area of interest, even though microelectronics, for example, is used in the creation of new media products. The challenge, therefore, is finding firms that make a realistic contribution in both areas and are not merely using pre-formed content or pre-existing computation (hardware or software) solutions. As we found in our study, it is necessary to have a fairly relaxed attitude to what constitutes a "contribution" in order to have a sufficient number of firms to study.

While all the above definitions are helpful, they fail to reflect the new ways in which people engage with these technologies and the implications for their future use. There is a further dimension to understanding the new media industry. New media technology offers the possibility of "two-way" engagement that did not previously exist. This potential for interactivity was stressed by Rice and Associates who define new media as "those communication technologies, typically involving computer capabilities that allow or facilitate interactivity among users or between users and information" (1984, p. 35). This modifies our definition to the extent that we intend new media to be understood as holding the potential for such interactivity, in contrast to the existing media that only send messages, such as radio, television, and newspapers.

Several of the industry players we interviewed were concerned about the perception of new media as cultural industry and not an economic industry. They thought new media should be seen as an industry that needs private

FIGURE 1
New Media Defined as a Continuum

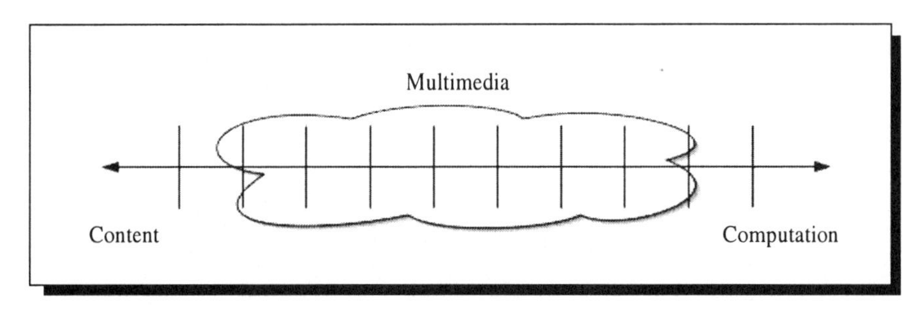

investment rather than public support; in contrast to the way that government usually views cultural industry. According to these respondents, money needs to be invested in every facet of the new media industry. Examples include the need to build large staging areas for post-production work and to finance content that does not fit within the Canadian Content rules required of government programs.

One interviewee remarked, "we can't expect a kid with a grant to produce a cultural CD-ROM product about the fur trade in Canada to turn around and launch a new media company." The change in the funding structure for an evolving industry like new media is dependent on a shift in the definition of new media from a cultural industry to an industrial sector.

THE NEW MEDIA INDUSTRY CLUSTER IN BC

The debates over definition reflect the fact that the industry is changing rapidly. Studies of the new media industry will therefore reflect a young and fast-moving business environment. This is certainly the case with two recent, major studies of the new media industry in BC, used for the purposes of this chapter.

The New Media BC (NMBC) Association conducted a report in consultation with government, industry stakeholders, and the Centre for Policy Research on Science and Technology (CPROST) in the spring of 2003. New Media BC (NMBC) is a non-profit industry association serving BC multimedia companies. Located in downtown Vancouver, NMBC was founded in 1998 and currently has a membership of more than 135 companies (Simmons 2003) and is sponsored by the Western Economic Diversification Fund, a federal agency. Taking a quantitative approach the NMBC survey was comprised of 32 questions that were posted on the New Media BC Web site. An invitation to visit the link was e-mailed to more than 700 new media companies, almost 40 percent of which responded, representing more than 260 companies. Survey participants were asked to provide information about their company's age, geographic location, core activities, target markets, human resources, innovative activities, financing and ownership, revenues, export activity, collaboration, industry challenges, and reasons for remaining in British Columbia. Responses in each section were then cross-tabulated by age, sector, revenues, and location. While some of the responses simply corroborated findings from previous industry surveys, others revealed significant change and growth in the industry during the past five years.

Carried out under the auspices of the Innovation Systems Research Network (ISRN) our study is the second recent attempt to better understand the new media industry in BC. In contrast to NMBC's quantitative methods, our study focuses on a qualitative approach using a detailed questionnaire to conduct in-depth interviews. An extensive literature review was used to help build up a picture of the industry as a whole. To date, we have completed approximately 70 interviews: 51 with companies and individuals and 19 interviews with representatives from the industry associations and the government agencies.

The NMBC study identified a geographic clustering of companies in the BC Lower Mainland, centred on certain areas of Greater Vancouver: Yaletown, Gastown, and central Burnaby. The industry is still in its infancy: over half of the companies (61.8 percent) surveyed have been in operation for less than six years. This figure compares to earlier statistics from a study conducted in 2000 by PriceWaterhouseCoopers in which 54 percent of new media companies were found to be four years old or younger. Today's new media industry in BC is comprised of more than 700 companies and provides full-time employment for approximately 14,000 people. This is a significant increase from 1998 when PriceWaterhouseCoopers reported that new media employed 1,800 people. Although 79 percent of BC's multimedia companies are located within the Lower Mainland, the industry is beginning to flourish in other areas of the province such as the Okanagan and Kootenay regions and Vancouver Island. The boom period of the late 1990s is reflected in the age of the companies. Nearly one-third of the companies started in that period and more than half (62.9 percent) are less than six years old. The typical Lower Mainland company is slightly older than its counterparts in the rest of the province. A majority of BC's multimedia companies, 87 percent, are privately owned, 8.8 percent are public companies, 2.3 percent are subsidiaries of Canadian parent companies and 1.5 percent are subsidiaries of foreign parents. Regardless of the location, the majority of new media companies, 52.5 percent, employ less than six people. The statistics show that as companies mature their employee count also increases (NewMediaBC 2003).

The NMBC survey divided the multimedia industry into three subcategories according to function: content providers, enablers, and delivery. Content providers are those companies that develop digital content such as games, animation, Web design/development, e-learning, visual effects, streaming media, and Internet publishing. Deliverers provide the "pipelines" for digital content and include Web hosts, Internet service-providers, and telecommunications companies. Enablers provide the tools and resources necessary for the delivery of digital content such as software development, Internet applications, digital compression technologies, security software, and e-commerce applications. In contrast, we distinguish between firms according to products and markets. Our interviews and discussions revealed that rather than being a single cluster, the industry in BC is composed of four different sub-clusters. Although they carry out some similar functions, they are distinguished by their markets and products. In essence the sub-clusters correspond to different new media product areas.

E-learning Sub-cluster

Companies in this sector specialize in producing and developing content and tools to facilitate online learning. It is the youngest sub-cluster in BC's new media industry and most of its companies are very small, often one-man shops operating out of home offices. BC's first e-learning companies developed in response to a need for distance learning. E-Learning BC, a volunteer-based association for the e-learning industry started in April 2000, currently has 70 active member companies throughout the province. Initially operating under the New Media Association, E-Learning BC separated from NMBC in early 2002 and formed a separate association. E-Learning BC actively markets its members and their services within Canada and internationally. They also provide opportunities for their members to collaborate. The biggest challenge in the industry, however, is that the small size and limited resources of most companies impedes their ability to reach international markets. Exactly how much support the government should provide to the e-learning companies remains a point of contention. Some interviewees suggested that the government should provide minimal support and allow the "cream of the crop" to rise to the top and avoid creating an artificial economic environment. Others argue that because other jurisdictions are directly supporting their new media industries, BC and Canada need to do the same.

The Gaming Sub-cluster

Don Mattrick, President of the Worldwide Studios at Electronic Arts, played an important role in the formation of Vancouver's gaming industry. The success of his first game, *Evolution*, allowed him and his former partner, Jeff Sember, to start their own company, Distinctive Software, in the early 1980s. The firm grew to 80 people and in 1991 it was bought by Electronic Arts (EA). EA continued to grow and by 2003, its revenues had reached $2.5 billion. Today, the company has offices in BC, the United States, and Europe and develops, publishes, and distributes software worldwide for video game systems, personal computers, and the Internet. As Distinctive Software and later EA evolved, a number of employees chose to start their own companies, including Radical Entertainment, founded by Ian Wilkinson and Rory Armes, and Relic Entertainment, founded by Alex Garden. These companies, in turn, spun out more companies, such as Barking Dog Studios and Black Box Games, which contributed to the vibrancy of Vancouver's multimedia industry. EA has since bought Black Box Games (in June 2002) to gain downtown presence in Vancouver. The majority of companies in this space develop computer games for various platforms such as PlayStation®2 computer entertainment system, the PlayStation®, Xbox™ video game console from Microsoft, the Nintendo GameCube™, the Game Boy® Advanced and PCs, which they then sell to publishers.

The Animation Sub-cluster

The first BC-based animation companies were founded in the late 1980s, providing service work for major Hollywood companies and large animation companies such as Nelvana and Cinar, in Toronto and Montreal. The first companies, Bardel Entertainment and Studio B, gave rise to more animation companies and the sub-cluster grew through the 1990s. The nature of the activities evolved from doing service work for the US-based companies to producing indigenous content through co-production with other Canadian and international companies. Producing and selling their own content is important to animation companies because it complements revenue, adds value to the company and contributes to a portfolio that can be showcased to prospective clients. In 1994, several key players formed the Association of British Columbia Animation Producers (ABCAP) to voice the concerns of the sector. In 2001, ABCAP conducted a survey of some 60 firms either directly working in or deriving some portion of their business from the global animation industry. According to the survey, in fiscal year 2000, the revenues of the sector exceeded $460 million, a significant increase from $27 million in 1991 and $286 million in 1998 (ABCAP 2001). Today ABCAP has 30 member companies, most of them located in the Lower Mainland.

Web Services Sub-cluster

This final sub-cluster is different from the other three in that its companies engage in very diverse activities. Their main products are a combination of services such as the preparation of various materials for e-commerce, Internet marketing, and the development of transport technologies. Some of the pioneers in the transactional aspects of the Internet were located in Vancouver. The majority of these companies are small and privately owned, some working on a contract basis. Most of the companies operate locally but some, such as Blast Radius, have achieved a worldwide reputation, and have a portfolio of clients that includes some of the world's largest firms: Nike, Nintendo, and Heineken.

Characteristics of the New Media Cluster in BC

The nature of the new media concentration or cluster in BC is best described by four key factors: innovation, financing, the marketplace, location, and what we have called the cluster milieu. These are each discussed below.

Innovation

Innovation is important in any industry because it serves as an indicator of future growth and profitability. This is particularly so with new media (and indeed in any high-tech industry) where business is about intangible, intellectual products, as opposed to tangible, physical products. Most of our interviewees reported that during the last three years their company had developed new products and services and/or made significant improvements to their current products and services. Many said that a company needs to constantly innovate to stay competitive. The following quote is representative of the innovation strategy laid out by the majority of companies: "Yes. We're in the technology business. If we did not [innovate], we would not be around" (Asgari 2002). The NMBC survey used the production of intellectual property (IP) as the key measure for assessing the innovative climate in the cluster. The survey asked participants whether their company produces its own IP and the majority of survey participants (80.5 percent) indicated that they did. The survey also asked participants to specify what kind of IP the company produces. The majority of the IP takes the form of products (66.8 percent), for example, hardware, software, and copyright materials such as photographs or artwork. Significantly less is produced in the form of trademarks (23.6 percent), patented inventions (14.4 percent) and industrial design (13.5 percent). Within the overall picture, we found great complexity, as companies move fluidly, sometimes within one transaction, between the different sub-functions of enabler, service provider, and delivery (NewMediaBC 2003).

Financing

Financing is a key issue for the development of a cluster. Ready access to capital, either from private or public sources, is the cornerstone for fostering growth and the developing of new ideas. Conversely, the lack of investors and available capital has been identified as a key inhibitor of growth. The NMBC survey asked respondents about the current financing of their companies, the types of funding and percentage of total funding that each type supplied. Most of the companies (83 percent) that responded to this question indicated that they have received funding from their founders or proprietors. This is by far the most common form of funding in the industry. In comparison, the next most common form of funding is debt (19 percent). Almost half of respondents (48 percent) who receive funding from their founders/proprietors reported that this source accounts for 100 percent of their total funding (NewMediaBC 2003).

Notoriously, during the euphoria of the dot-com era, the wide availability of funding sometimes led to over-enthusiastic, if not unwise investment choices. Investors worldwide suffered heavy losses when the dot-com bubble burst. In the aftermath, investment strategies in BC for technology-based industries have become more conservative. Companies founded in or before early 2001 were able to obtain their first round of financing relatively easily, but the situation has changed dramatically since then. "November 2000 [first round] was relatively easy ... however, our second round ... we signed at noon on September 11, 2001, so you can imagine that was challenging ... we're trying to hold off right now on going back to raise money and hunkering down to conserve our capital" (Calvin 2002).

The three most common sources of funding are: founders/proprietors (23 percent), government (22 percent), and venture capital (19 percent).

FIGURE 2
The Percentage of Funding Received Across all Companies

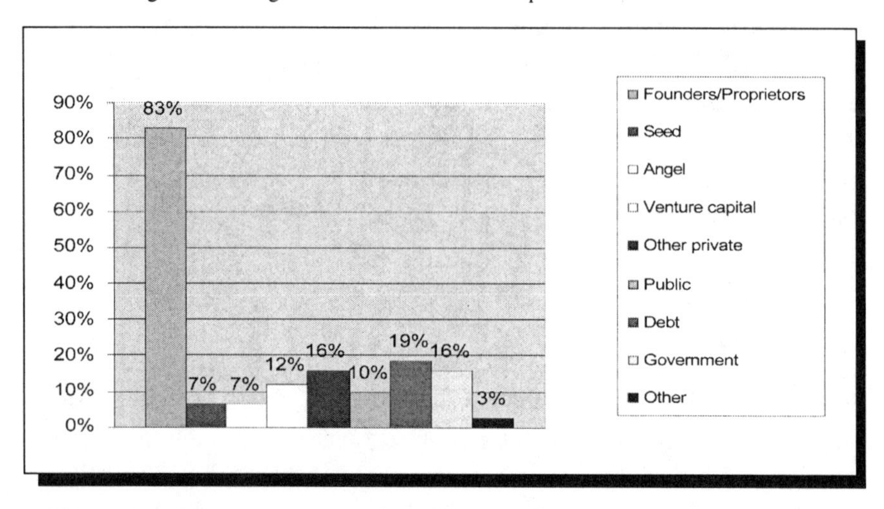

Note: Total companies answering: 217. Respondents were allowed to choose more than one answer, therefore percentages do not add to 100 percent.

Although funding from government and venture capital were listed as the second and third most common funding sources, they were also ranked as the top two most difficult sources to access. While government funding may be seen as a reflection of confidence in the industry, it is vulnerable to time lags. This is particularly detrimental to a fast-moving, technology-based industry such as new media.

The most common types of government funding received by respondents (see Table 1) are those from the Industrial Research Assistance Program (IRAP) (15 percent of respondents) and the Scientific Research and Experimental Development (SRED) tax credit system (13 percent of respondents). It should be noted that these two programs are highly agent-driven, well-connected and embedded in the social network of the region. For instance, one of the most popular workshops in the region is the regularly held SRED seminar on how to gain government funding and deal with tax issues. As a result, the firms pursuing these as funding sources have to engage with the networks in the region. In the process, they are gaining expertise and becoming more linked with the cluster.

TABLE 1
Government Sources of Funding

Choice	*Count*	*Percent of Answered*
Industrial Research Assistance Program (IRAP)	33	15.00
Scientific Research and Experimental Development	28	13.00
Advanced Systems Institute (ASI)	8	4.00
Technology Partnership Canada	1	0.40
Precommercialization Assistance (PA)	2	0.90
Community Futures	11	5.00
Science Council of BC	6	3.00
Telefilm Canada New Media Fund	14	6.00
Canadian Heritage	9	4.00
Program for Export Market Development (PEMD)	3	1.00
CANARIE	10	5.00
Other (please specify)	16	7.00
None	148	66.00

Notes: Total companies answering: 223. Respondents were allowed to choose more than one answer, therefore percentages do not add to 100 percent.

Despite the relative importance of public sources of funding, the companies had a mixed response to the impact of government programs on the development of the new media cluster. Some made the point that because government funding often favours products, many new media companies which are service-focused do not qualify.

Export Activities and Available Markets

It is important for any industry to protect itself against adverse local economic conditions by maximizing its export potential and expanding it revenue base. According to the NMBC survey, more than 75 percent of BC's multimedia companies currently export products, services or both. Seventy-six percent of respondents plan to increase or expand their export activities in the next 18 months. Canada and the United States are the most important markets, with the United Kingdom ranked a distant third. Approximately 43 percent of companies generate less than 10 percent of their revenue through export activities. At the other end of the scale, however, 30 percent of exporting companies (mostly larger companies) generate more than 50 percent of their income from exports (NewMediaBC 2003).

Over half of all survey respondents (55 percent) reported that their companies attend trade shows. This activity is important for companies of all ages, regions, sizes, and sectors because it is the place for meeting potential clients and showcasing the company's work. Over 155 different shows were named with a great diversity of focus and location. Some of the more frequently named shows were Comdex, E3 (Electronic Entertainment Expo), Banff Film and TV Festival, World Education Market, and CES (Consumer Electronic Show).

Location, Location, Location

Vancouver is famous for its lifestyle, and many of the people employed in the new media industry readily admit that they are drawn here for the advantages the city has to offer. NMBC asked survey participants to rank the importance of the seven factors to their company's staying in BC and to their company's growth (lifestyle, large talent pool, access to western US, low dollar, strong sense of community in new media, and access to the Asian markets). The top three reasons for staying in BC were lifestyle, availability of large talent pool, and access to the western US (Pacific Northwest and California). The same trends were identified during our interviews.

A mild climate, proximity to the ocean and the mountains, and orientation toward healthy living foster a creative environment and attract people who like to "work and play" where they live. According to one respondent "we are located in Kits[1] because we find that the region offers a balance between work and leisure that appeals to our staff and clients. Vancouver itself, as a location, offers many business opportunities for smaller companies to interact with larger companies. There is competition but there is also a lot of trust from clients" (E-MediaDesign 2001).

Vancouver has several educational institutions that supply a high-quality workforce for the cluster. Many companies fill entry positions by hiring local graduates from Simon Fraser University, Vancouver Film School, British Columbia Institute of Technology and UBC: "I would say that looking at our staff they're a pretty unique bunch ... Vancouver and the surrounding area has a lot of institutions that offer fantastic training programs like the universities and colleges and there's often workshops as well as simply self-training and gaining insights on trends from other companies and colleagues and friends" (Dave Olsson 2001).

When asked whether the local labour market possesses distinctive or unique sets of skills, knowledge or capabilities, many interviewees indicated that creativity, flexibility in terms of services and products, and relevant industry experience are notable attributes in the Vancouver labour market. "People here are balanced between having a very strong work ethic but also a very strong sense of creativity" (Act360Media 2001). The president of a medium-sized company, who worked in Canada and the US, identified a strong work ethic as an important quality of the local workforce: "The work ethic is a lot stronger here [in Canada] in an interesting way. They are good about creating a workaholic workforce and Canadians are more selfless in terms of contributing to the business ... in the US people are more focused on their own careers and values ... Canadians in this perspective are more balanced" (Cotichini 2003). Multiculturalism was also mentioned by a number of interviewees: "Because Vancouver is a multicultural society, which has different industries and clusters, it provides a varied source of skilled workers" (Faber 2001).

When asked about employees who had left their company within the last three years, interviewees identified three reasons: downsizing, personality fit, and contractors. As companies go through different stages in financing, they have to adjust their budget and downsize; however, they often end up hiring the same employees back if a new wave of financing is set. Also, many companies have a very strong culture and new employees sometimes find it difficult to fit in. When asked how many downsized employees are hired by other firms within the region, the interviewees claimed that the majority were able find work locally.

Finally, due to the project-based environment, some companies prefer to hire contract employees or/and consultants: "Animation is a global and nomadic industry; most of the people that work in it move; unless you've been working for 20 years in Nirvana. Most of the employees are hired on a contract basis except the core group and they like it ... they travel and they bring experience. It is the lifestyle choice" (Ward 2003). Experienced consultants provide knowledge-transfer and promote an innovative climate within the industry: "People like myself ... who have all those skills ... there are small companies that need help in getting started, with market research, introductions ... and we help them. It is important to have floaters like me ... we know everybody" (Haman 2002).

Responses to the question of finding replacements for key employees were varied. The majority stated that employers have an advantage in today's market environment: "There is a very large pool of knowledge workers here. The last technical support position we advertised for we received a little over 250 applications" (Edis 2002). Other interviewees said that it would be difficult to find replacements due to the specific set of skills required and/or the personality fit: "It would not be easy to replace key people and so we work hard to keep them here. It is the culture – their contribution to it and their understanding of our business, they are passionate about what we are doing. When you lose that passion, it is hard to replace it" (Cotichini 2003).

Since many companies have clients in the US, especially in California and Silicon Valley, they see the advantage of being close to the US market, while paying salaries in Canadian dollars: "Blast Radius's business model is a near shore development model: we are almost as cheap as India (relative to cost of living) and lot easier to work with. We also provide high quality work" (Fergunsson 2003). Sharing a time zone with California and the Silicon Valley was also seen as advantageous by a number of companies, especially those that specialize in game development: "It is pretty easy [to fly to San Francisco] ... I will go down to California in the morning and will be home and see the kids for dinner" (Daniel 2002).

Cluster Milieu

The NMBC survey revealed a high degree of collaboration among BC new media companies, which suggests that this activity is vital to industry growth and sustainability. Many companies actively collaborate with other BC or Canadian businesses, international organizations, universities or colleges, and research institutions. The survey showed that 40 percent have collaborated with at least one company within the province; 18 percent of the companies collaborate with other companies in Canada; 20 percent collaborate with foreign companies; and 15 percent collaborate with postsecondary institutions. The average number of collaborative partners within BC was slightly greater than five (NewMediaBC 2003). The strong sense of community was ranked as "somewhat important" in the new media industry and this corresponds with the statistics that find 44.9 percent of companies located in the Lower Mainland and 43.9 percent in the rest of BC.

Industry associations play a vital role in supporting collaboration among the new media companies. Many interviewees described them as a "supporting network" for the industry. Industry associations provide various activities and events to help their members meet potential investors and partners, discuss ideas, find work, and learn the latest trends. One of the interviewees, who is an active member of E-Learning BC, pointed out that to serve the e-learning market it is important to have: "a network of supporting services to offer a full range of services to these clients, so what E-learning BC made possible is for these people to actually get to know each other and build a level of trust with them, so you can be in a client situation, see an opportunity and immediately react to it, knowing what you are recommending and you could in fact go so far as to take responsibility and subcontract and know that you are not putting your own reputation at risk, and that's very important" (Stewart 2003).

ANALYSIS

Together with the NMBC study our work provides a full picture of the current new media industry in BC. The findings indicate some of the necessary ingredients for an industrial cluster in the multimedia sector. There are a reasonably large number of firms and they are growing. There is policy awareness of the sector and regulatory (tax and other) initiatives to promote firms. Finally, the industry itself has sufficient self-organization to have created several networking events and associations. Other aspects of our work differ from common perceptions about clusters: the youth of the industry, the turnover of companies, the presence of four sub-clusters, the speed at which the industry moves. There is more to a cluster, however, than just the outward appearance. There is also the question of whether firms actually work together. Our study raises a number of interesting questions. Can we describe this as a cluster, albeit a young, immature and fast moving one, or is this in fact a different phenomenon? This final section explores some of these questions and attempts to reach some conclusions.

According to generally accepted definitions of an industrial cluster, for the multimedia industry in Vancouver to qualify as a cluster it would have to have certain "commonalities and complementarities" (Porter 1998*a*). Making an operational definition of commonalities and complementarities proved relatively straightforward. For commonalities we looked for common membership in industry associations, common "root" firms or investors in the genealogy of firms (Smith 2002), and attendance at industry-specific social and networking functions. Complementarities were defined according to the "value-chain" model of local inter-firm connections, using the horizontal and vertical linkages (Mytelka and Farinelli 2000) and Porter's own "value-chain" model (Porter 1985).

Commonalities were assessed, in this preliminary formulation, according to a simple "high/low" descriptive variable, with the common features of firms measured relative to the industry group as a whole. Complementarities were assessed similarly, with firms measured as having low complementarities if they reported few local value-chain linkages and high if they reported several local linkages. These firm-level measures were later aggregated for the sub-industry groupings that make up the cluster as a whole, although again, these results should be considered very preliminary at this point.

A two-dimensional grid was created out of these two aspects of clustering, with commonalities on the vertical axis and complementarities on the horizontal axis (see Figure 3). The location of the various sub-sectors in Figure 3 are subject to revision as we learn more about the use of this method of understanding clusters. For now, it is illustrative, at best. That said, anecdotal evidence from the case studies provides some background for why we might place firms in one quadrant or the other.

FIGURE 3
Commonalities and Complementarities as Cluster Metrics

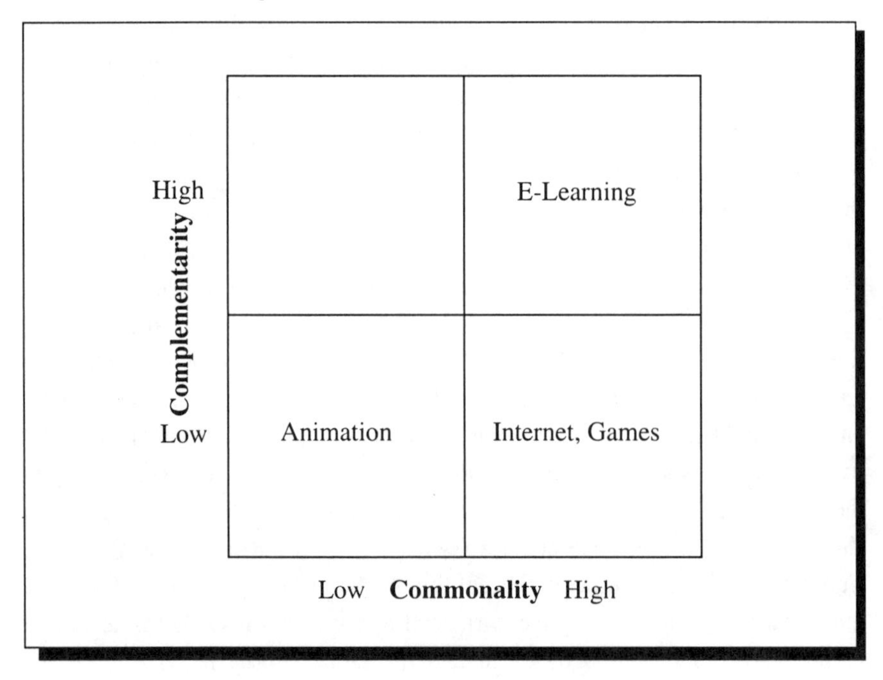

The animation subgroup within the multimedia cluster scores low on the commonality measure in large part because there are relatively few firms engaged in this business in Vancouver. As a result, they are more or less unique in their approach and business practices and technology. They score low on the complementarity axis because they are relatively unconnected to each other, albeit well-connected to the global marketplace and supply of animation talent. The exception to this is the moderate connections that they have to local schools and design programs.

The Internet (Web design, streaming media) and electronic games firms are relatively more numerous and more homogenous in their approach, technology, and customer profile. They are extremely well-connected to each other in industry associations and networking events, and they have deep connections into the local higher education scene. They do not have a pronounced value-chain orientation, however, and therefore score relatively low on the complementarity axis. The one exception to this is Electronic Arts, which is sufficiently large to have begun purchasing specialized services from related firms such as sound studios. This may be a prelude to a more integrated local cluster.

The e-learning group is comprised of a diverse set of firms with a high degree of commonality as evidenced by their common roots (many spinoffs from local university research groups or related firms), their extensive networking activities (there is a special interest group on e-learning within the NewMediaBC association, for example), and their collective initiatives in the areas of marketing and promotion. At the same time, the industry is more vertically linked than some of the other subgroups, with specialized firms taking up roles in service of other local firms, and some firms contracting with others to fill in gaps in their technology, market, and business practices.

The grid is useful in a number of ways. It provides a visual interpretation of some of the findings and it increases confidence in the idea of the new media industry in BC as a *bona fide* cluster phenomenon. It reinforces our conclusion about the existence of the sub-clusters by showing that they are distinct and differentiated and also how they may relate to each other. The grid is constructed according to measurable quantitative data, however, and as such, it does not take into account more intangible findings, nor does it allow us to critically evaluate the cluster as a whole.

To allow us to take a more in-depth look, we have combined the Wolfe and Gertler (2003) indicators with our environment-knowledge-society factors in an evaluation below. In terms of inflows, the studies show all three forms identified by Wolfe and Gertler: capital, people, and knowledge. Where outflows are concerned, the studies report that more than three-quarters of companies export products or services, and that these are more than likely to be bound up with intellectual property concerns. It would seem that in those terms, this sector meets the criteria.

The New Media Industry Environment

It is important to consider the nature of the environment in which the new media sector is embedded, namely, the new media industry, both in BC and worldwide. Our discussion of the industry reflects just how new the industry is and how rapidly it is changing. In many respects, the marketplace is still being created, so it is difficult to say much about products, services, competitors, and niches that will still be true in 12 months time. What can be said is that this is an extremely dynamic environment. Companies entering it have to "hit the ground running" and be prepared to be flexible. This is a growing market, so what does this mean for the development of a cluster? The new media marketplace represents a vast potential of possibilities to be explored. Smaller companies, as commonly found in a new industry, cannot afford to explore the market possibilities on their own. It is far more effective to do so in collaboration with others because they can cover more "ground" by sharing knowledge and experience. Clusters develop because of the efficacy of such networking behaviours for the companies. Is this efficacy reflected in the networking mechanisms and behaviours of the new media companies in BC?

If we consider the level of knowledge-related activity in this sector, the section above on innovation reflects how significant R&D and the production of IP are in this sector. It is more difficult to establish the presence of functions and mechanisms that enable varied knowledge search and creation, such as the levels of diversity between firms and the ratio of exploring to exploiting firms.

In the absence of direct research to establish the levels of such activities, we can only speculate based on available studies. Our study identified the presence of firms that acted in some respect as "incubators," that provide training or apprenticeship opportunities for individuals to learn skills and gain experience. Once people achieve a necessary level of expertise with the incubator firms, they move on to larger firms where they can command higher wages. The role of incubator firms is crucial in underpinning the value creation of more successful firms. Each sub-cluster certainly appears to be engaged both in R&D and in the delivery of products and services. Companies also exhibit subtle differences. Does this mean that they are able to cover enough ground to enable the cluster to be sustainable over time? This is a question of critical mass that will only be resolved through time.

Of equal significance to these issues are mechanisms that enable knowledge or information to move through the cluster and to be shared. These are strongly linked to the movement of individuals between companies, as the studies indicate that it is a common occurrence that both freelancers and other individuals move from company to company. Such movement is an indication of an environment in which people are aware of expertise and experience, and know each other outside the boundaries of their own firms. It also helps to foster some sense of belonging, albeit a tenuous one. The presence of collaboration among competing firms also supports the evidence that the industry operates as a cluster. Our interviews suggest that firms readily engaged in collaborative relationships with both competitors and the supporting network, such as industry associations. Finally, a significant dimension of social factors is the presence of shared attitudes and value systems, which is certainly indicated by the studies, as many respondents reflected similar views about what attracted them to Vancouver, and others made reference to a unique type of work ethic.

CLUSTER OR WHIRLWIND?

From this brief analysis it certainly would appear that some of the mechanisms that enable a cluster to flourish are present among new media firms in BC. What is not clear is whether they are sufficient. The whole picture is complicated by the speed of movement of the industry itself, and the proximity to a large and powerful market in the US. The final indicator identified by Wolfe and Gertler (2003), historical path dependence, is concerned with the evidence of history. As the statistics above demonstrate, the BC industry is characterized by a fast turnover of companies. Overall, firms are fairly young and small. This may indicate that the sector is just moving too fast for a cluster to have time to develop. The proximity of the US may diffuse the attention of BC firms by causing them to focus on the US market. A further consideration of historical issues must be mentioned. The literature on clusters holds that they are not overnight phenomena — they take several decades to develop. In some cases, the cluster is the underlying socio-economic phenomenon in the context of successive technological waves. For instance, as Figure 4 shows, Silicon

Valley has gone through several such waves and is now considering what the next step should be. It is not clear whether this is also the case here in BC with the existing group of companies.

FIGURE 4
The Evolution of Silicon Valley, 1950–2000

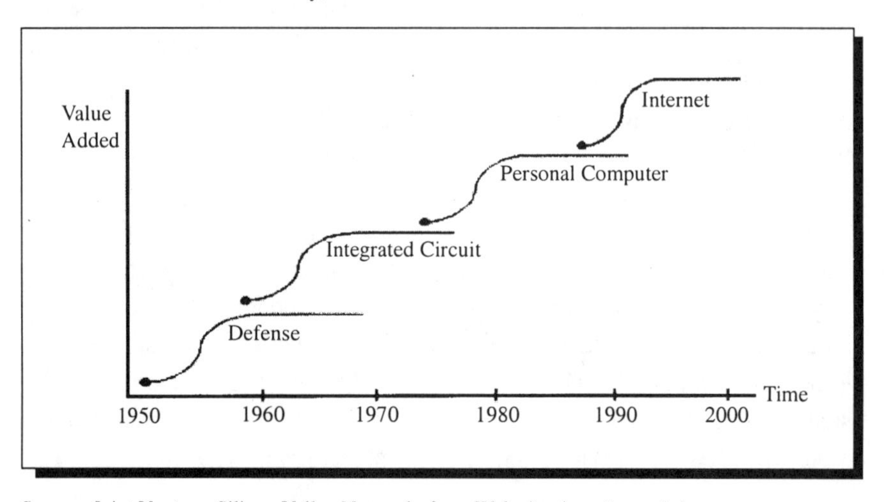

Source: Joint Venture: Silicon Valley Network, from Web site: http://www.jointventure.org

The new media sector in BC exhibits some characteristics of a cluster, but it is not clear if it is developing into a cluster with enduring potential. It is possible that in fact it is not a cluster at all. We propose two possible explanations for the current phenomenon.

- This is a cluster, but one that we are fortunate enough to study during its very birth, with all the attendant difficulties and problems it is encountering. This process is further complicated by the speed of the industry in which it is operating.
- This is not a cluster, although it does have many cluster characteristics. It is a whirlwind of activity focused on the technological excitement of new media industry. It has had perforce to develop some cluster dimensions, but in fact, it does not contain sufficient momentum or cohesion to ever become a cluster.

Is it possible to know which of these best defines the activities and the structure of the new media industry in BC? While we do not know how the evolving new media industry in BC is best defined, we contend that this short analysis is able to examine an evolving phenomenon that is not documented elsewhere in the literature.

NOTE

[1]Kitsilano — a "trendy" neighbourhood in Vancouver, famous for its beaches, restaurants and shopping.

REFERENCES

Association of British Columbia Animators and Illustrators (ABCAP). 2001. *Frame by Frame BC's Animation Industry*. Vancouver: ABCAP.

Act360Media. 2001. Personal interview with M. Petrusevich. Vancouver.

Amin A. 2000. "Industrial Districts," in *A Companion to Economic Geography*, ed. E. Sheppard and T. Barnes. Oxford: Blackwell.

Asgari, M. P. N. I. 2002. Personal interview with D. Wolstenholme. Vancouver.

Axelsson B. and G. Easton. 1992. *Industrial Networks — A New View of Reality*. London: Routledge.

Becattini, G. 1990. "The Marshallian Industrial District as a Socio-Economic Notion," in *Industrial Districts and Inter-Firm Co-operation in Italy*, ed. F. Pyke *et al.* Geneva: International Institute for Labour Studies.

Braczyk, H.-J., G. Fuchs and H.-G. Wolf, eds. 1999. *Multimedia and Regional Economic Restructuring*. London: Routledge.

Calvin, P.M. 2002. Personal interview with D. Wolstenholme. Vancouver.

Castells, M. 2001. *The Internet Galaxy: Reflections on the Internet, Business, and Society*. New York: Oxford University Press.

Coase, R. 1937. "The Nature of the Firm," *Economica* 4:386-405.

Cooke, P. 1999. "New Media and New Economy Cluster Dynamics," in *Multimedia and Regional Economic Restructuring,* ed. Braczyk, Fuchs and Wolf.

Coro, G. and R. Grandinetti. 1999. "Evolutionary Patterns of Italian District Networks," *Human Systems Management* 18(2):117-29.

Cotichini, C.M.T. 2003. Personal interview with M. Petrusevich. Vancouver.

Daniel, M. R. E. 2002. Personal interview with R.P. Smith. Vancouver.

Dave Olsson, C.F. 2001. Personal interview with K. Warfield. Vancouver.

Dimou, P. 1994. "The Industrial District: A Stage of a Diffuse Industrialisation Process — The Case of Roanne," *European Planning Studies* 1(2):23-38.

Edis, D.I.T. 2002. Personal interview with D. Wolstenholme. Vancouver.

E-MediaDesign. 2001. Personal interview with K. Warfield. Vancouver.

Faber, H.B.B. 2001. Personal interview with M. Petrusevich. Vancouver.

Fergunsson, M.B.R. 2003. Personal interview with M. Petrusevich. Vancouver.

Garofoli, G. 1991. "Local Networks, Innovation and Policy in Italian Industrial Districts," in *Regions Reconsidered*, ed. E.M. Bergman *et al.* London: Mansell.

Gilder, G. 1993. "Telecosm – 'The New Rule of Wireless,'" *Forbes Supplement*, pp. 96-111.

Haman, A. 2002. Personal interview with R.P. Smith. Vancouver.

Housser, S. and Vancouver Public Library. 2001. *New Media Directory* (3rd ed.). Vancouver, BC: Vancouver Public Library.

Lievrouw, L.A. and S. Livingstone, eds. 2002. *The Handbook of New Media.* London: Sage Publications.

Marshall, A. 1890. *Principles of Economics.* London: Macmillan.

Mytelka, L. and F. Farinelli. 2000. "Local Clusters, Innovation Systems and Sustained Competitiveness." Discussion Paper Series, No. 2000-5. New York: The United Nations University, Institute for New Technologies.

NewMediaBC. 2003. *New Media in BC 2003 Industry Survey.* Vancouver: New Media BC: 1-63.

Nohria, N. and R.G. Eccles, eds. 1992. *Networks and Organisations — Structure, Form and Action.* Boston: Harvard Business School Press.

Olson, J.E. 1988. "Customers, Competition, and Compatibility: A New Convergence Shaping the Information Age," *International Journal of Technology Management* 3(4):375-80.

Piore, M.J. and C.F. Sabel. 1984. *The Second Industrial Divide – Possibilities for Prosperity.* New York: Basic Books Inc.

Porter, M.E. 1985. *Competitive Advantage: Creating and Sustaining Superior Performance.* New York: Free Press.

—— 1991. *Canada at the Crossroads: The Reality of a New Competitive Environment.* Ottawa: Business Council on National Issues and Supply and Services Canada.

—— 1998*a*. "Clusters and the New Economics of Competition," *Harvard Business Review* 76 (November-December):77-90.

—— 1998*b*. *On Competition.* Cambridge, MA: Harvard Business School Publishing.

Pyke, F. and W. Sengenberger, eds. 1992. *Industrial Districts and Local Economic Regeneration.* Geneva: International Institute for Labour Studies.

Reddy, W.M. 1984. *The Rise of Market Culture.* New York: Cambridge University Press.

Rice, R.E. and Associates, eds.1984. *The New Media: Communication, Research and Technology.* Beverly Hills, CA: Sage.

Saxenian, A. 1990. "Regional Networks and the Resurgence of Silicon Valley," *Californian Management Review* 33(1):89-112.

—— 1994. *Regional Advantage: Culture and Competition in Silicon Valley and Route 128.* Cambridge, MA: Harvard University Press.

Schiavon, F. 2003. Personal interview with M. Petrusevich. Vancouver.

Sengenberger, W. and F. Pyke. 1992. "Industrial Districts and Local Economic Regeneration: Research and Policy Issues," in *Industrial Districts and Local Economic Regeneration*, ed. Pyke and Sengenberger.

Simmons, K. 2003. Personal interview with M. Petrusevich. Vancouver.

Smith, R. 2002. "Techmaps: A Tool for Understanding Social Capital for Technological Innovation at a Regional Level," in *Innovation and Entrepreneurship in Western Canada: From Family Businesses to Multinationals*, ed. J.J. Chrisman, J.A.D. Holbrook and J.H. Chua. Calgary: University of Calgary Press.

Stewart, B. A. T. S. S. G. 2003. Personal interview with M. Petrusevich. Vancouver.

Ward, B. 2003. Personal interview with M. Petrusevich. Vancouver.

Wardrip-Fruin, N. and N. Montfort, eds. 2003. *The New Media Reader.* Cambridge, MA: The MIT Press.

Webster, S. 2003. Personal interview with M. Petrusevich. Vancouver.

Williamson, O.E. 1975. *Markets and Hierarchies: Analysis and Anti-Trust Implications.* New York: Free Press.

Wolfe, D.A. and M.S. Gertler. 2003. "Clusters Old and New: Lessons from the ISRN Study of Cluster Development," in *Clusters Old and New: The Transition to a Knowledge Economy in Canada's Regions*, ed. D.A. Wolfe. Kingston and Montreal: School of Policy Studies, Queen's University and McGill-Queen's University Press.

CONTRIBUTORS

JOHN N.H. BRITTON is Professor of Geography at the University of Toronto where he has been a faculty member since 1968. His research has probed a number of issues in Canadian economic development, including the implications of foreign ownership and the impacts of trade liberalization. Recently, he used survey data from Toronto's electronics cluster to explore a variety of network relationships which are important for high technology firms. In particular, in his latest published papers he has focused on the importance of knowledge inputs obtained through extra-regional networks.

CATHERINE CHEVRIER is a research assistant on the Innovations Systems Research Network project and a Master's of Business Administration candidate at University of Québec in Montreal.

NICOLA CROWDEN has a BA (Hons) in Criminology and was awarded her MA in Communication in 2003 from Simon Fraser University. She has been a research assistant at the Centre for Policy Research on Science and Technology and the project officer for the Women's Advisory Group on Innovation Studies at Simon Fraser University.

SUSAN FITZGIBBON is a PhD candidate in the Department of Geography at Queen's University. She has co-authored papers in the field of population health and demography, with a particular interest in the geography of labour. She is currently working on two research projects examining changes in the structure and organization of the automotive products industry in Canada.

HAELI GOERTZEN received her MA in Political Economy at Carleton University after completing her BSc in Chemistry. Her research interests include the political economy of agriculture and science and technology policy.

J. ADAM HOLBROOK is Adjunct Professor and Associate Director of the Centre for Policy Research on Science and Technology (CPROST) at Simon Fraser University. Prof. Holbrook was trained as a physicist and electrical engineer and is a registered professional engineer in the provinces of Ontario and British Columbia. At CPROST his research activities centre on the analysis of science, technology, and innovation activities in both the public sector and the private sector. He is the leader of InnoCom, a network of researchers in innovation and industrial cluster studies in western Canada and is a member of the management committee of the national Innovation Systems Research Network.

JOHN HOLMES is Professor and Head of Geography and a Faculty Associate in the School of Industrial Relations at Queen's University. He is the author of numerous journal articles and chapters on economic geography with a particular focus on the automobile industry. His current research is on the dynamics of restructuring and industrial competition in the Canadian automotive components industry and union responses to continental economic integration, restructuring and work reorganization in the broader North American automobile industry. He is also a member of two research networks concerned with employment, work, and institutional innovation in the new economy.

MÉLANIE KÉROACK has an MBA in International Business from Laval University and the University of Monterrey Mexico. She holds a specialization in Finance (Albany New York) and Marketing (Collège Ste-Foy, Quebec). In October 2002, she joined Dr. Landry's team with a mandate to study the optical cluster in Quebec City. She has written a paper on the challenges of the internationalization process for firms in the optical industry and has worked in the optical field as a marketing analyst.

PRADEEP KUMAR is a Professor in the School of Industrial Relations at Queen's University. His research focuses on the goals, strategies and performance of Canada's unions. Currently he is engaged (with Gregor Murray, Université de Montréal) in studies of union renewal based on surveys and case studies of innovations and change in labour organizations, as well as research on the dynamics of restructuring and industrial competition in the Canadian automotive components industry (with John Holmes). In addition to numerous academic articles on various aspects of industrial relations, he has published several books: *From Uniformity to Divergence: Industrial Relations in Canada and the United States* (1993), and *Unions and Workplace Change in Canada* (1995). He is a member of a number of research networks.

RÉJEAN LANDRY is the holder of a Chair on Knowledge Transfer and Innovation funded by the Canadian Health Services Research Foundation and the Canadian Institute of Health Research. Dr. Landry is Professor at the Department of Management of the Faculty of Business at Laval University where he lectures on knowledge transfer. He has published extensively on public policies, innovation and knowledge transfer. His most recent works on knowledge transfer have been published in *Public Administration Review*, *Technological Forecasting and Social Change*, *Research Policy*, *Technovation and Science Communication*.

GERRY LEGARE is a PhD candidate in the Department of Geography and Planning at the University of Toronto and a lecturer in the Department of Geography at Brock University. His doctoral work examines the structure and organization of the film and television industry in Toronto, with a specific focus on cluster properties, labour market organization, and innovation.

MATTHEW LUCAS is a PhD candidate in the Institute for the History and Philosophy of Science and Technology and a research associate at the Program on Globalisation and Regional Innovation Systems at the Centre for International Studies, both at the University of Toronto. His doctoral work examines university-industry research collaboration. He is currently preparing a comparison of ICT clusters across Canada.

JANE MCCARTHY is Visiting Faculty at the Centre for Policy Research on Science and Technology at Simon Fraser University and Visiting Fellow at Cranfield School of Management. Her central research interests are innovation networks and clusters. Her doctoral thesis presented a complex systems-based model of industrial network adaptability.

CAROLAN MCLARNEY is Associate Professor of Strategic Management and International Business at Dalhousie University. Currently Dr. McLarney is the Director of the Centre for International Business Studies at Dalhousie. Prior to completing a PhD at York University, she held management positions in various companies in the hospital, transportation, and consulting sectors.

LYNN K. MYTELKA is Director of the United Nations University Institute for New Technologies (UNU-INTECH) in Maastricht, The Netherlands. She is also a Distinguished Research Professor at Carleton University and Honorary Professor in Development Economics at the University of Maastricht.

MATHIEU OUIMET is a PhD candidate and researcher at the CHSRF/CIHR Chair on Knowledge Transfer and Innovation at the Business School at Laval University. He was awarded a Policy Research Prize from the Canadian Policy Research Initiative of the Government of Canada in 2000 and has been selected for the dean's list of the Graduate School of Laval University. In the last three years, he has co-authored a book on social network analysis and many articles on research transfer and innovation.

MICHELLE PETRUSEVICH is an MA candidate in the School of Communication at Simon Fraser University, and has been working as a research assistant at the Centre for Policy Research on Science and Technology (CPROST) since 2002. She has contributed to the studies of the Vancouver new media cluster as a student researcher for the Innovation Systems Research Network and the New Media BC association. Her research interests include social capital, new media, and social networks.

SHAWNA REIBLING is a research associate at the Centre for Sustainable Communities Canada and a Master's candidate at CPROST, School of Communication, Simon Fraser University. Her most recent work includes a water technology cluster analysis and coordination of a conference on Environmentally Sustainable Technology Innovations. Her research interests include innovation policy studies, the open source movement, and gender issues. Her thesis work focuses on technology and social responsibility.

PHILIP ROSSON holds the Killam Chair of Technology, Innovation and Marketing in the School of Business Administration at Dalhousie University. Between 1999 and 2002, he served as co-editor of the *Canadian Journal of Administrative Sciences*. He has published widely, with a special emphasis on the growth strategies of small and medium-sized companies, particularly in foreign markets. Dr. Rosson was educated in England where he earned his MA (Lancaster) and PhD (Bath) degrees.

SERGE ROUSSEAU is a research assistant on the Innovations Systems Research Network project. He is also an MBA candidate at the University of Québec in Montreal.

TOD RUTHERFORD is Associate Professor in the Department of Geography in the Maxwell School of Citizenship and Public Affairs at Syracuse University. His research interests include labour relations and innovation, the restructuring of work and buyer-supplier relations in the automobile industry, and labour market policy.

MONICA SALAZAR is a PhD candidate at Simon Fraser University and a researcher at the Centre for Policy Research on Science and Technology at Simon Fraser University. Ms. Salazar has a BSc in Economics from Universidad Rosario in Bogota, Colombia and an MSc in Technical Change and Industrial Strategy from the University of Manchester (UK). Her professional career has included positions at the Colombian Institute for the Development of Science and Technology and the Colombian Department of National Planning.

RICHARD SMITH is Associate Professor of Communication at Simon Fraser University. He is also Director of the Centre for Policy Research on Science and Technology (CPROST). Smith's research focus is new media as a technology, as a business, and as a factor in social change. He has an ongoing interest in technology for education, surveillance of public spaces, online communities, and the wireless information society.

DIANE-GABRIELLE TREMBLAY has a Canada Research Chair on the socio-organizational challenges of the knowledge economy and is co chair of the Bell-Téluq-Enap Chair on Technology and Work organization. She is Professor of labour economics and human resources management, as well as Director of Research at Télé-université, Université du Québec. She has published many articles and books, amongst which are textbooks on labour economics and sociology of work; two books on working time issues, and a book on local economic development. She is a member of the executive committee of the Sociology of Work research committee of the International Sociological

Association, as well as of the executive of the Society for the Advancement of Socio-Economics. She has a PhD in Economics from the Université de la Sorbonne in Paris.

KATIE WARFIELD received her Master's of Communication and Certificate in Urban Planning from Simon Fraser University in 2003. At SFU she studied multicultural cityscapes with a focus on multicultural urban planning issues in Vancouver.

NAOMI WEINER is currently completing a Bachelor of Arts in Communication and Sociology at Simon Fraser University. While pursuing her degree she has been an assistant research manager for a retail development consultant (Thomas Consultants Inc.) and a research assistant at the Centre for Policy research on Science and Technology at Simon Fraser University.

DAVID A. WOLFE is Professor of Political Science at the University of Toronto and Co-Director (with Meric Gertler) of the Program on Globalization and Regional Innovation Systems (PROGRIS) at the Centre for International Studies, which serves as the national secretariat for the Innovation Systems Research Network. He is National Coordinator of the ISRN and principal investigator on its Major Collaborative Research Initiative on Innovation Systems and Economic Development: The Role of Local and Regional Clusters in Canada. Recent publications include *Innovation, Institutions and Territory: Regional Innovation Systems in Canada, Clusters and Regional Innovation: Economic Development in Canada* (both co-edited with J. Adam Holbrook), *Clusters Old and New: The Transition to a Knowledge Economy in Canada's Regions,* and *Innovation and Social Learning: Institutional Adaptation in an Era of Technological Change* (co-edited with Meric S. Gertler).

Queen's Policy Studies
Recent Publications

The Queen's Policy Studies Series is dedicated to the exploration of major policy issues that confront governments in Canada and other western nations. McGill-Queen's University Press is the exclusive world representative and distributor of books in the series.

School of Policy Studies

Canada Without Armed Forces? Douglas L. Bland (ed.), 2004
Paper ISBN 1-55339-036-9 Cloth 1-55339-037-7

Campaigns for International Security: Canada's Defence Policy at the Turn of the Century, Douglas L. Bland and Sean M. Maloney, 2004
Paper ISBN 0-88911-962-7 Cloth 0-88911-964-3

Understanding Innovation in Canadian Industry, Fred Gault (ed.), 2003
Paper ISBN 1-55339-030-X Cloth ISBN 1-55339-031-8

Delicate Dances: Public Policy and the Nonprofit Sector, Kathy L. Brock (ed.), 2003
Paper ISBN 0-88911-953-8 Cloth ISBN 0-88911-955-4

Beyond the National Divide: Regional Dimensions of Industrial Relations, Mark Thompson, Joseph B. Rose and Anthony E. Smith (eds.), 2003
Paper ISBN 0-88911-963-5 Cloth ISBN 0-88911-965-1

The Nonprofit Sector in Interesting Times: Case Studies in a Changing Sector, Kathy L. Brock and Keith G. Banting (eds.), 2003
Paper ISBN 0-88911-941-4 Cloth ISBN 0-88911-943-0

Clusters Old and New: The Transition to a Knowledge Economy in Canada's Regions, David A. Wolfe (ed.), 2003 Paper ISBN 0-88911-959-7 Cloth ISBN 0-88911-961-9

The e-Connected World: Risks and Opportunities, Stephen Coleman (ed.), 2003
Paper ISBN 0-88911-945-7 Cloth ISBN 0-88911-947-3

Knowledge, Clusters and Regional Innovation: Economic Development in Canada, J. Adam Holbrook and David A. Wolfe (eds.), 2002
Paper ISBN 0-88911-919-8 Cloth ISBN 0-88911-917-1

Lessons of Everyday Law/Le droit du quotidien, Roderick Alexander Macdonald, 2002
Paper ISBN 0-88911-915-5 Cloth ISBN 0-88911-913-9

Improving Connections Between Governments and Nonprofit and Voluntary Organizations: Public Policy and the Third Sector, Kathy L. Brock (ed.), 2002
Paper ISBN 0-88911-899-X Cloth ISBN 0-88911-907-4

Institute of Intergovernmental Relations

Canada: The State of the Federation 2002, vol. 16, *Reconsidering the Institutions of Canadian Federalism,* J. Peter Meekison, Hamish Telford and Harvey Lazar (eds.), 2004
Paper ISBN 1-55339-009-1 Cloth ISBN 1-55339-008-3

Federalism and Labour Market Policy: Comparing Different Governance and Employment Strategies, Alain Noël (ed.), 2004
Paper ISBN 1-55339-006-7 Cloth ISBN 1-55339-007-5

The Impact of Global and Regional Integration on Federal Systems: A Comparative Analysis, Harvey Lazar, Hamish Telford and Ronald L. Watts (eds.), 2003
Paper ISBN 1-55339-002-4 Cloth ISBN 1-55339-003-2

Canada: The State of the Federation 2001, vol. 15, *Canadian Political Culture(s) in Transition,* Hamish Telford and Harvey Lazar (eds.), 2002
Paper ISBN 0-88911-863-9 Cloth ISBN 0-88911-851-5

Federalism, Democracy and Disability Policy in Canada, Alan Puttee (ed.), 2002
Paper ISBN 0-88911-855-8 Cloth ISBN 1-55339-001-6, ISBN 0-88911-845-0 (set)

Comparaison des régimes fédéraux, 2e éd., Ronald L. Watts, 2002
ISBN 1-55339-005-9

John Deutsch Institute for the Study of Economic Policy

The 2003 Federal Budget: Conflicting Tensions, Charles M. Beach and Thomas A. Wilson (eds.), 2004 Paper ISBN 0-88911-958-9 Cloth ISBN 0-88911-956-2

Canadian Immigration Policy for the 21st Century, Charles M. Beach, Alan G. Green and Jeffrey G. Reitz (eds.), 2003 Paper ISBN 0-88911-954-6 Cloth ISBN 0-88911-952-X

Framing Financial Structure in an Information Environment, Thomas J. Courchene and Edwin H. Neave (eds.), Policy Forum Series no. 38, 2003
Paper ISBN 0-88911-950-3 Cloth ISBN 0-88911-948-1

Towards Evidence-Based Policy for Canadian Education/Vers des politiques canadiennes d'éducation fondées sur la recherche, Patrice de Broucker and/et Arthur Sweetman (eds./dirs.), 2002 Paper ISBN 0-88911-946-5 Cloth ISBN 0-88911-944-9

Money, Markets and Mobility: Celebrating the Ideas of Robert A. Mundell, Nobel Laureate in Economic Sciences, Thomas J. Courchene (ed.), 2002
Paper ISBN 0-88911-820-5 Cloth ISBN 0-88911-818-3

Available from: McGill-Queen's University Press
c/o Georgetown Terminal Warehouses
34 Armstrong Avenue
Georgetown, Ontario L7G 4R9
Tel: (877) 864-8477
Fax: (877) 864-4272
E-mail: orders@gtwcanada.com